Forest and Plant Bioproducts Division 2019

Held at the 2019 AIChE Annual Meeting

Orlando, Florida, USA
10-15 November 2019

ISBN: 978-1-7138-0539-7

Printed from e-media with permission by:

Curran Associates, Inc.
57 Morehouse Lane
Red Hook, NY 12571

proceedings.com

Some format issues inherent in the e-media version may also appear in this print version.

Printed with permission by Curran Associates, Inc. (2020)

For permission requests, please contact AIChE
at the address below.

AIChE
120 Wall Street, FL 23
New York, NY 10005-4020

Phone: (800) 242-4363
Fax: (203) 775-5177

www.aiche.org

Additional copies of this publication are available from:

Curran Associates, Inc.
57 Morehouse Lane
Red Hook, NY 12571 USA
Phone: 845-758-0400
Fax: 845-758-2633
Email: curran@proceedings.com
Web: www.proceedings.com

Forest and Plant Bioproducts Division 2019

Held at the 2019 AIChE Annual Meeting

Orlando, Florida, USA
10-15 November 2019

AIChE
The Global Home of Chemical Engineers

TABLE OF CONTENTS

574007 Kraft Lignin Color Reduction for Potential UV Absorbent Applications in Transparent and UV Protection Films

The dark brown color of kraft lignin (KL) restricts its potential utilization as structural material or functional nanofiller in engineering plastics and composites. Lignin in its native state in wood is nearly colorless. However, it is transformed into the dark brown colored material during kraft pulping process. Several efforts are discussed in literature related to lignin color reduction; however, most of them degrade the lignin aromatic structure reducing its UV absorption properties. The goal of this research is to improve the scope of kraft lignin as an UV absorbent nanomaterial in transparent and UV protecting polymer composite films. This will be carried out through its modification for reducing color chromophores without significantly degrading the lignin aromatic structure. The current study will investigate the effect of peroxide and chlorite bleaching of kraft lignin with and without blocking of phenolic hydroxyl groups on its optical properties. Phenolic hydroxyl groups can be blocked using acetylation or carboxymethylation of KL, where latter eliminates the use of organic solvent during bleaching step due to the solubility carboxymethylated KL in aquoues medium.The preliminary results of the acetylated kraft lignin (Ac-KL) and peroxide bleached kraft lignin (PB-KL) showed substantial increase in their visible light transmission without significantly affecting UV blocking properties (Figure.1). 0.1 wt.% solution of Ac-KL and PB-KL in dioxane/water (0.5/0.5 v/v) showed ~160% and ~130% increase in their visible light (550 nm) transmission compared to unmodified kraft lignin. Moreover, the UV absorptivity of the Ac-KL and PB-KL did not change significantly as seen from their UV-Vis spectra and extinction coefficient (Ɛ) measured at 255 nm. Acetylation of KL replaced its phenolic hydroxyl groups with the acetyl groups which reduced the lignin color by preventing the formation of quinone and quinone methide chromophores. Whereas, color reduction in peroxide bleaching of KL is attributed to the oxidation of existing chromophores to colorless bodies. In order to obtain maximum color reduction, two-stage modification will be employed wherein acetylated/carboxymethylated lignin will be oxidized using peroxide bleaching. Similarly, these modified lignin will also be treated using chlorite bleaching. These approaches can synergistically reduce the lignin color by reducing the chromophore forming side reactions of unmodified lignin during bleaching and through the oxidation of the existing chromophores to colorless bodies. Finally, transparent, UV protection films will be processed through the casting of two-stage modified KL and polyvinyl alcohol aqueous mixtures. The resultant films will be characterized for their optical, thermal and mechanical properties.

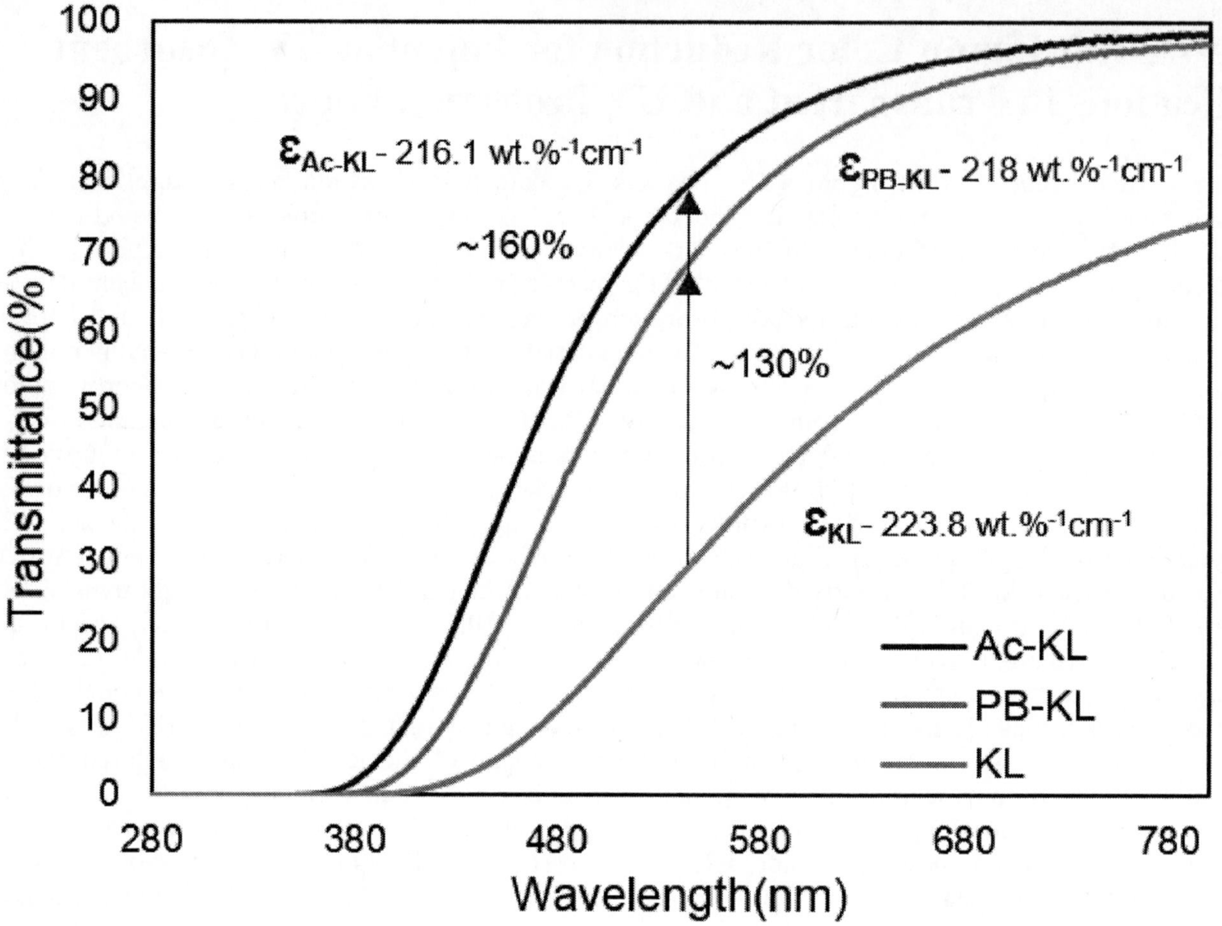

Figure 1. UV-Vis spectra of the kraft lignin samples

578365 Multistream Integrated Biorefinery Achieved By Integration of Lignin Bioconversion and Carbon Fiber Manufacturing

Lignin is the main compound of lignocellulosic biomass and the major waste in paper industry. Recent technology advancements have enabled the bioconversion of lignin to higher value products. In addition, lignin has long been sought after as a feedstock for renewable carbon fiber. Our recent study discovered that lignin carbon fiber quality is defined by molecular uniformity, and lignin bioconversion can remove the small molecular weight lignin to derive more uniform fraction. These discoveries have enabled us to develop a novel integrated process to derive multiple products out of lignocellulosic biorefinery. The lignin-containing biorefinery was used for producing lipid and PHA with *Rhodococcus opacus* and *Pseudomonas putida*, respectively. The lignin after fermentation was precipitated to manufacture carbon fiber. The carbon fiber derived from lignin after fermentation has 66.2% higher tensile strength and 34.4% higher modulus of elasticity than the lignin from original biorefinery waste. This finding highlighted a new multistream biorefinery to produce both lipid or PHA from bioconversion and higher quality carbon fiber. The multistream biorefinery will improve the sustainability and economics of biorefinery substantially.

577679 Enhancing Fertility of Marginal Soils Using Lignin-Rich Hydrolysis Residues

During lignocellulosic biofuel production, cell wall sugars are converted into fuel. However, residual biomass remains following enzymatic hydrolysis that is enriched in lignin and high in organic carbon and nitrogen. This residue could be used as a cost effective and environmentally friendly soil enhancement. Soils with higher organic carbon content have shown increased water holding capacity and pH. It is also possible for the residual nitrogen to function as a fertilizer, reducing the need for other nutrient amendments. The goal of this experiment was to determine the effect of corn stover lignin residue amendments on the characteristics of maize grown in marginal soil. Onaway-Ossineke fine sandy loam was collected from the MSU Forest Biomass Innovation Center (FBIC) in Escanaba, MI. Lignin residue was generated from enzymatic hydrolysis of ammonia fiber expansion (AFEX) pretreated corn stover. Residue was mixed with soil in amounts that increased the percentage of organic carbon by 0%, 2%, 10%, or 25% and packed in triplicate soil columns. Water was added weekly to the columns to match weekly average precipitation in Houghton, MI. The filtrate was evaluated for pH, volume, and ammonium and nitrate contents. The initial and final CN content of the soil columns was evaluated to determine stability of added carbon and nitrogen. To investigate the effect of amendment addition on maize growth, soils were amended with residue in the same amounts, and added to 3.8 L pots in six replicates for each treatment. Fertilizers were mixed into the soil at amounts equal to 125 lb P_2O_5 and 200 lb K_2O/acre. Urea was also added to half the replicates to maintain a constant nitrogen loading of 230 lb N/acre. Prior to planting, corn seeds were sorted to ensure consistent mass, sterilized, and planted at 1" depth in pots in triplicate. Once 2" in height, the plants were thinned to a single plant per pot and evaluated over time for growth, ear production, and nitrogen deficiency. Initial work showed that the addition of lignin residue had no major effect on pH, which ranged from 8.3 – 8.9 for all levels of amendment, and no significant effect on water holding capacity. Final CN content of the soils was analyzed and showed a decrease in carbon and nitrogen for all soil columns over time, regardless of the level of amendment. Maize greenhouse experiments are currently underway to evaluate the potential for residue nitrogen to improve plant growth.

581345 Evaluation of Pennycress Biochar for Adsorption of Nitrophenols from Water

Field pennycress (*Thlaspi arvense L.*) is a winter cover crop that has been growing in popularity as a means to increase profitability for soybean farmers in the Midwestern United States. Pennycress seeds can be used to produce biodiesel and jet fuel. The oil extraction from the seeds leaves copious amounts of pressed cake. In this work, pennycress press cake was examined as a potential feedstock for biochar production that could be used as filter media for the separation of toxic organic compounds from drinking water. Pennycress pressed cake is carbonized to obtain pennycress biochar. The biochar is activated using hydrochloric acid and characterized for its elemental composition, surface area, surface functional moieties (FT-IR), crystallinity (XRD), and surface morphology (SEM). The biochar has a surface area of 283 m^2/g and a bulk density of 1.84 g/cm^3. The biochar is evaluated for adsorption of USEPA listed priority pollutants 2-, 3-, and 4-nitrophenols from the water. Results of equilibrium adsorption and column breakthrough studies will be presented. This study helps in assessing the overall use of pennycress plant products and the economic benefits of growing pennycress as a cover crop. The study also provides an option for treating drinking water as well as information for separation of isomeric mononitrophenols.

576494 Surface Tension Analysis of Lignin Nanoparticles and Cellulose Nanocrystals with Magnetic Properties in Water-Ethanol Mixtures Under Varying Experimental Conditions

The surface tension of ethanol-water mixtures varies significantly as a function of concentration; however, the ethanol concentration in certain processes, such as broth fermentations, is limited to low alcohol percentages. In an effort to find a practical and alternative approach to surfactants use to decrease surface tension, this work analyzes the use of lignin nanoparticles (LNPs) and cellulose-nanocrystals (CNCs) combined with magnetic nanoparticles for controlling surface tension in water-ethanol mixtures. The addition of iron oxide (Fe_3O_4) to LNPs and CNCs allowed the generation of hybrid bio-based nanocomposites which are able to be manipulated under the presence of an external magnetic field.

The synthesis of hybrid LNPs and CNCs with iron oxide nanoparticles was performed using co-precipitation techniques, and these efforts are currently being reported by the authors elsewhere. Briefly, LNPs/Fe_3O_4 were synthesized using a modified co-precipitation procedure with two steps. The first step used iron chloride (II) and iron chloride (III) salts in the presence of a base to produce Fe_3O_4 nanoparticles. After several washes with ultra-pure type I water, a second step in the preparation was done by adding lignin solubilized in basic solution at low concentration via a self-assembly method using a syringe pump at a fast addition rate of 30 mL/min. The second bio-based hybrid nanoparticles, CNC/Fe_3O_4, were synthesized using the same iron salts but in a one-step co-precipitation method with the CNCs already dispersed in the solution. Vibrating Sampling Magnetometer (VSM) and Raman spectroscopy techniques were used to obtain the magnetic and structural properties of both nanocomposites.

In this study, the specific aim is to evaluate the structure-property relationships on the surface tension of water-ethanol mixtures caused by the addition of LNPs/Fe_3O_4 and CNC/Fe_3O_4 at varying concentrations, temperature, and magnetic fields. A goniometer instrument was used to collect surface tension measurements, and a custom 3-D printed holder was used to manipulating the magnetic field by adding cylindrical-shaped earth magnets while the measurements were recorded. Various studies were performed to analyze the effects on surface tension caused by the nanocomposites. The first experiment analyzed the impact of at least six different water-ethanol mixtures with each hybrid nanocomposite, LNPs/Fe_3O_4 and CNC/Fe_3O_4, dispersed in a concentration varying from 0 – 2.5 wt%. There was little to no change in surface tension (~2 mN/m) at low nanoparticles concentrations (< 1 wt%) for both bio-based nanocomposites. However, for the LNPs/Fe_3O_4 at concentrations above 1 wt%, there was a significant decrease in surface tension at low ethanol concentrations in water-ethanol mixtures (10 % or less). To analyze the effect of temperature conditions, the LNP/ Fe_3O_4 concentration was kept constant at 1 wt% in a pure water solution between 20 °C - 35°C. The surface tension decreased with increased temperature for all samples. Results revealed a maximum reduction of ~8.5 mN/m for the LNP/ Fe_3O_4 at 35 °C. Lastly, the magnetic field effects on surface tension were analyzed at various LNP/ Fe_3O_4 concentration in water at 5.22, 7.05, and 9.24 Gauss, respectively. On the contrary to the effects of increasing temperature, the surface tension increased as a function of the applied magnetic field and LNP/ Fe_3O_4 concentration. The change in surface tension was greater for higher magnetic fields. Overall, this study shows that the surface tension of water-ethanol mixtures can be increased or decreased by using LNP/ Fe_3O_4 or CNC/Fe_3O_4 under different experimental conditions, including nanoparticle concentration, temperature and applied magnetic field. These findings using bio-based magnetic nanocomposites to control surface tension could result in enhancements for ethanol extraction from aqueous-based solutions using currently-available chemical separation techniques.

581460 Enhanced Flame Retardancy of Poly(butylene terephthalate) with Biobased Flame Retardant

The development of halogen-free flame retardants has recently received increasing attention as consumer health issues have been associated with conventional halogen-based flame retardants. Biobased flame retardants are popular due to their environmental friendliness. During this study, we produced a biobased flame retardant from phosphorylated vegetable oil, synthesized by a ring-opening reaction in epoxidized linseed oil. The structure and degree of phosphorylation of the linseed oil were characterized with Nuclear Magnetic Resonance (NMR) spectroscopy. The synthesized biobased flame retardant was blended with an engineering plastic, poly(butylene terephthalate) (PBT), and the flammability of the blends were investigated with the UL-94 vertical and horizontal flame tests. It was discovered that the fire class of PBT changed from non-rating to V-2 at only 7.5 wt% loading of the synthesized flame retardant. Increasing the content of modified vegetable oil further increased the flame retardancy to a V-0 fire class. The fire mechanism of the blends was studied with Thermogravimetric Analysis coupled to Fourier Transform Infrared Spectroscopy (TGA-FTIR). The developed biobased flame retardant exhibits high potential to replace conventional halogen-based flame retardants.

Acknowledgement: The financial support from the Ontario Ministry of Agriculture, Food and Rural Affairs (OMAFRA)/University of Guelph - Bioeconomy for Industrial Uses Research Program (Project # 030251); and the Agriculture and Agri-Food Canada (AAFC) and IGPC Ethanol, Canada through Bioindustrial Innovation Canada (BIC) Bioproducts AgSci Cluster Program (Project # 053786 and 054015).

566259 Efficient Conversion of Pulp and Paper Sludge to Cellulose Nanomaterials and Their Application As Building Blocks for Constructing Conductive Nanopapers and Hydrogels

Recently, cellulose nanomaterials (CNM) include cellulose nanocrystals (CNCs) and cellulose nanofibrils (CNFs) produced from cellulosic materials (e.g. wood, agricultural and industrial waste, etc.) have received increasing attention due to their unique properties such as high specific surface area, high elastic modulus, high thermal stability, as well as renewability and biodegradability. These unique properties make CNM as promising building blocks for the manufacture of many sustainable nanomaterials.

In this work, we firstly tried to prepare CNFs from pulp and paper sludge (PPS) by a sustainable process of formic acid (FA) hydrolysis pretreatment combined with microfluidization. It was found that the mild FA hydrolysis (at 95 ºC for 3-6 h) pretreatment could hydrolyze most of hemicellulose, swell and break down the cellulose fibers, and the cellulosic solid residue (CSR) with a high yield (over 75%) could be obtained. The resultant CSR could be further converted to CNFs with relatively low-intensity microfluidization (only two passes). Then, we tried to introduce a small amount of $FeCl_3$ into the FA hydrolysis process as catalyst to enhance the hydrolysis efficiency. Results showed that CNCs could be produced with a comparable yield of 20-30 wt.% with traditional sulfuric acid hydrolysis, and the CSR could be further fibrillated to CNFs with low-intensity fibrillation. In addition, FA could be readily recovered and reused with the recovery rate higher than 90%. Intriguingly, both the resultant CNCs and CNFs showed good dispersibility in DMSO, DMF and DMAC respectively because of the introduction of ester groups on the surface of the products during FA hydrolysis.

Furthermore, cellulose nanopapers (CNP) were prepared from the above obtained CNFs dispersion in DMAC by a vacuum filtration approach. A thin layer of silver nanowires could be further introduced on top of the CNP surface to make conductive CNP. Results showed that the resultant conductive CNP had high transparency (over 80% at 550 nm), high thermal stability (onset thermal degradation temperature was over 320 ºC), low resistance (45.57 O/sq) and excellent water-resistance (could remain the shape in water over 100 days), which showed great promise for the development of flexible and multi-functional electronics (e.g. touch screen).

In addition, both of the CNCs and CNFs were coated with a thin layer of polypyrrole (PPy) via in situ polymerization. The PPy-coated CNCs and CNFs as conductive nanofillers were further incorporated into polyvinyl alcohol (PVA)-borax hydrogels respectively. It was found that the composite hydrogels showed high viscoelasticity and mechanical strength, as well as good conductivity, which had potential applications in multifunctional smart soft materials.

568106 Rheological Characterization of Nanocellulose Materials for Quality Control

Nanocellulose material is a renewable and sustainable nanomaterial produced from abundant cellulose sources. Its high strength and biodegradability make it attractive to many applications such as composites, coatings and rheological modifiers. Two types of nanocelluloses are mostly studied in the literature. One is cellulose nanocrystal (CNC), which are rod-like colloidal particles. The other is TEMPO oxidized cellulose nanofibril (TEMPO-CNF), which are flexible fibers. To ensure consistent production of high quality nanocellulose materials, one of the most urgent issues to be addressed is the lack of standardized, rapid and reliable characterization methods in the manufacturing environment. Electron microscopy and light scattering are two current characterization techniques. Nevertheless, the former is expensive and time-consuming, and the latter is extremely sensitive to aggregation. Moreover, both techniques only probe a small portion of the nanocellulose sample, which may misrepresent the sample's bulk properties. Rheology provides a fast and cost-effective way to characterize nanocelluloses in large volume. In this work, we will show the potential of rheology as a quality control tool for nanocellulose materials. We developed detailed test protocols to obtain reliable viscosity data. We studied the influence of concentration on rheological properties of CNC and TEMPO-CNF, and formulated a rheological model to accurately capture the viscosity across shear rates and concentrations. We will demonstrate the model's capability to determine the concentration of a nanocellulose sample by measuring the sample's viscosity for both CNC and TEMPO-CNF. We will also show that the processing conditions, including the effect of pressure homogenization (pressure and number of passes) and the carboxylate content of TEMPO-CNF, can be identified through viscosity measurements. The measurement and the model can be applied on the industrial-grade viscometer and the capillary rheometer for quality control.

573466 Cellulose Nanofiber and Polypyrrole Based Conducting Composite Films with Improved Conductivity, Strength, Water Resistance and Electromagnetic Shielding Properties

Polypyrrole (PPy) is interesting conducting polymer due to its high conductivity, good environmental stability, ease of synthesis and non-toxic nature. It has applicability in several areas such as antistatic materials, electromagnetic shielding, biomedical devices, various biosensing and electronic devices as a substitute for metallic conductors or semiconductors. However, PPy is difficult to process unlike the conventional thermoplastic and thermoset polymers using methods such as melt processing and solution casting. It is insoluble in common solvents and decomposes before reaching its melting temperature. Several attempts were carried out in order to improve the processability of PPy using various polymer matrices such as rubber, polyamides, polyolefins and polyesters. Recently there has been increasing attention towards utilization of biobased materials such as cellulose nanofibers (CNF) as substrate for the preparation of flexible and conducting composite films. This approach is more sustainable, renewable and reduces the dependence on the petroleum based non-renewable resources. Most methods in literature for preparation CNF based conducting composite films have used in situ polymerization of pyrrole onto nanofibers followed by film preparation (coded here as ISF). ISF approach produces flexible conducting nanopaper with good conductivity but has poor mechanical strength due to loss of hydrogen bonding interaction between the nanofibers and high porosity of the film. Moreover, these films are not water resistant which further limits their utility in various electrochemical applications.

In this research we have developed a novel and simple approach for improving the electrical and physical properties of the CNF and PPy based composite films. Two approaches were employed, in first approach in situ polymerization of pyrrole was carried out on CNF film to obtain PPy/CNF composite films while in second approach polyvinyl alcohol (PVA) coated CNF film was used to obtain PPy/PVA-CNF composite films. PVA coated CNF film was prior heat treated at 150°C for 90 min to increase its water resistance. The SEM images of the resultant composite films revealed distinct difference in the surface morphology of these films (Figure 1). Highly porous and rough surface was observed for the ISF composite films, while in situ polymerized films had more uniform PPy coating compared to ISF. This has resulted in significant improvement in conductivity of PPy/CNF and PPy/PVA-CNF composite films (Table1). Higher conductivity in case of PPy/PVA-CNF than PPy/CNF was due to smoother and more uniform PPy coating resulted from the partially crosslinked PVA layer on CNF film and absorption of more monomer and oligomers inside the swollen PVA layer during in situ polymerization. Besides this, weaker interaction between partially crosslinked PVA and PPy would retain the good conducting state for PPy compared to CNF-PPy interphase. PPy/PVA-CNF and PPy/CNF also has higher mechanical strength compared to ISF composite films (Table 1). Tensile strength improvement for PPy/PVA-CNF and PPy/CNF compared to ISF films is due to reduced porosity and improved hydrogen bonding interaction between the nanofibers in these films. Moreover, the CNF/PPy and CNF/PVA-PPy films exhibited water stability due to hydrophobic and compact PPy coating while high porosity and poor interaction between PPy coated nanofibers in ISF films drastically reduced their water stability. Further, films were also characterized for their EMI shielding effectiveness which showed similar trend as conductivity with CNF/PVA-PPy films showing maximum shielding effectiveness (~15 dB). For a given thickness this is promising EMI shielding behavior compared to reported values in literature. These conducting composite films with improved performance have potential applications in the electronic devices, antistatic and anticorrosive coatings, intelligent clothes, sensors, flexible electrodes, and tissue engineering scaffolds.

Figure 1. SEM images showing surface morphology of composite films

Table 1. Physical and electrical properties of the CNF and CNF-PPy composite films

Sample	Conductivity (S/cm)	Tensile strength (Mpa)	EMI shielding effectiveness (dB)	Water stability (7 day mixing)
CNF	0	93.4	0.11	unstable
ISF	0.26	7.6	3.5	unstable
PPy/CNF	4.5	91.1	11.5	stable
PPy/PVA-CNF	9.6	103.1	14.5	stable

577210 Immobilization of Antibodies on Cellulose Nanocrystals for the Detection of Cancer Biomarkers

Cellulose nanocrystals (CNC) are an intriguing, sustainable material for biosensors due to their exceptional mechanical properties and nanoscale structure. However, relatively little is known about antibody immobilization on CNC; this is particularly true for the commonly available CNC that are produced by sulfuric acid extraction from woody biomass. This research used APTES-GA chemistry for immobilization of model antibodies on sulfonated CNC from the US Forest Products Lab. The following monoclonal antibodies were chosen for their use in cancer biomarker detection: alpha-fetoprotein (AFP) used in liver cancer detection, prostate-specific antigen (PSA) used in prostate cancer detection, and carcinoembryonic antigen (CEA) used in ovarian cancer detection. The results of immobilization of each of these biomarkers on solid CNC films and dispersed CNC were compared using thermogravimetric analysis (TGA), attenuated total reflection infrared spectroscopy (ATR-IR), and atomic force microscopy (AFM). Quartz crystal microbalance with dissipation monitoring (QCM-D) was used to follow each step of the immobilization in real time as well as to detect the antigen binding. In addition, QCM-D was used to quantify the amount of antibody immobilized as well as binding of the corresponding antigen. Both schemes resulted in antibody immobilization. Immobilization on the films was the simplest method, but greater immobilization and antigen binding were achieved on dispersed CNC. This is attributed to accessibility of the entire CNC's surface area in the dispersions. However, functionalization of dispersed CNC tended to result in CNC cross-linking and aggregation unless the reaction was performed at relatively dilute conditions. Furthermore, the presence of the APTES on the CNC surface inhibited the formation of highly aligned films. These results highlight that sulfonated CNC can be used for antibody immobilization and are an interesting green material for biosensors.

577374 High Shear Behavior of Nanocellulosics

Nanocellulosics, which are typically found in the form of Cellulose nanocrystals (CNCs) and Cellulose Nanofibrils (CNFs) provide promise as both filler and reinforcement materials. While much research is currently investigating potential applications of nanocellulosics, little is understood about the impact of high rate and stress, associated with many conventional processing techniques, on the rheology and structure of these materials. Our research investigates high shear rheology of nanocellulosics, highlighting impacts of temperature, concentration, and stress on the flow behavior and physical properties of these materials. We have characterized the potential degradation and relaxation characteristics of these materials. Furthermore, we have demonstrated the ability to generate master curves based on time-temperature and time-concentration superposition principles.

572480 Kinetic Modelling of Lignocellulose Hydrolysis at High Solids Loading

Biomass slurries, required to achieve relevant sugar concentrations for biofuel/bioproduct manufacture must approach 200 g/L or higher. As higher concentrations are achieved, the biomass slurries become shear thinning. Consequently, the manner in which lignocellulose is added to the reactor, mixing conditions, and sequence of enzyme addition combine to exert a major influence on formation of a “flowing” biomass suspension, and the rates and extents to which the cellulose is hydrolyzed. We report work for corn stover. Our preliminary data show that, in a mixed reactor, conversion at constant enzyme loading may remain constant as solids concentrations increase from 10 to 200 g/L. At low loadings and minimal mixing, mass transfer through free water is readily achieved, while at high solids shear stress interferes with the productivity of the reactor, since higher viscosity may also inhibit rates of hydrolysis. Mixing decreases apparent viscosity and increases mass transfer resulting in a constant conversion at high loadings. This work relates viscosity to conversion and proposes an approach to predict liquefaction and cellulose hydrolysis as a function of shear stress, mixing, and enzyme addition strategies.

569960 Fermentation of Xylose to Carboxylic Acids Under Low pH Conditions

Lignocellulosic biomass has been perceived to be a promising and widely available resource for conversion to bioethanol. Lignocellulosic materials contain both cellulose and hemicellulose in significant quantities which upon hydrolysis produces both C5 and C6 sugars. The inability of microorganisms to efficiently utilize xylose, the principal C5 sugar, results in significant sugar wastes from bioethanol production that impacts upon the economics of the cellulosic ethanol process. A biorefinery concept that promotes the comprehensive utilization of the hemicellulose and lignin component of biomass will favorably impact the economics of biomass to biofuel processes. In this work, the utilization of xylose in the production of lactic and acetic acids is explored using bacteria that can ferment xylose at pH ~ 4. Fed-batch fermentation studies have shown that ~ 67 g/L of lactic acid and 32 g/L of acetic acid can be produced from a feed containing ~ 80 g/L of xylose. The low pH of fermentation allows for the efficient extraction of the acids from the broth. Batch and continuous fermentation data will be presented along with a model that can predict acid productivities as a function of dilution rate.

570078 Comparison of Product Distribution, Content and Fermentability of Biomass in a Hybrid Thermochemical/Biological Processing Platform

Thermochemical processing is a promising method for the rapid depolymerization of biomass. This study investigated switchgrass, corn stover, red oak, hybrid poplar, and loblolly pine in terms of heteropolymer and elemental composition, and the distribution and composition of the fast pyrolysis products. Corn stover differed from other biomass types in that less of the biomass was recovered as sugar or phenolic oil (PO) and more of the biomass was recovered as bio-char and bio-gas. The sugar-rich aqueous stream recovered from the bio-oil heavy fraction was characterized in terms of sugar content and distribution, inhibitor content, and ability to support production of ethanol by *Escherichia coli* KO11+*lgk* as a model biorenewable product. Levoglucosan was the most abundant sugar from each type of biomass, followed by either xylose or cellobiosan. For hybrid poplar, cellobiosan accounted for 30 wt% of the total sugar pool. Each of the sugar streams also contained a variety of inhibitors, particularly 5-hydroxymethylfurfural (5-HMF) and methylcyclopentenolone. Methylcyclopentenolone, maple lactone, was found to decrease the specific growth rate of *E. coli* by 50% when present at 0.72 wt%, indicating that it is less toxic than furfural, acetic acid and guaiacol. Sugars produced from switchgrass contained 4-fold less contaminants on a per-sugar basis than those from poplar and pine. All of the sugar streams contained too many inhibitors to be used at an industrially feasible concentration without additional detoxification. The poplar-derived pyrolytic sugar syrup was particularly inhibitory, possibly due to the high abundance of aromatic hydrocarbons, such as xylenes, and anisoles.

576652 Superior Performance Biodiesel from Biomass Hydrolysate Fusel Alcohols and Bio-Oils: Fatty Acid Fusel Esters (FAFE) and Combustion Behavior

In an effort to address challenges related to fuel performance and feedstock limitation for biodiesel production, we demonstrated production of fatty acid fusel alcohol esters (FAFE) obtained by reacting fusel alcohols from bioconversion of mixed substrate biomass hydrolysates with oil. Firstly, we developed a microbial consortium-based biocatalysis strategy to utilize both the complex sugars and the amino acids in algae hydrolysates as a substrate for the production of C4 and C5 fusel alcohols. Secondly, we demonstrated the utilization of high free fatty acid containing bio-oil sources for enzymatic transformation in to fatty acid fusel esters (FAFE) using lipase from *Aspergillus oryzae*. The effect of different mole ratios of fusel alcohol mixture (isobutanol, 3-methyl-1-butanol and 2-methyl-1-butanol) on the transesterification of glyceryl trioleate and oleic acid was studied. Optimized reaction conditions achieved >97% yield of fatty acid conversion to FAFE. Thirdly, mixed ester surrogates of different oil sources such as *Chlamodomonas* biomass (green microalgae biomass), waste cooking oil, and *Rhodosporidium* biomass (lipidic yeast) were evaluated for use as a fuel for diesel engines. The data indicated that FAFE have superior combustion and cold-flow properties, a 34.6% shorter ignition delay time (leading to higher cetane values) and a 9∘C lower no-flow point in comparison to FAMEs.

565621 Fenton-Based Oxidation As a Pre-Treatment to Sugarcane Vinasse to Lower Inhibition to the Growth of Oleaginous Fungi

The utilization of sugarcane vinasse as a feasible culture medium for fungi is hindered by the presence of inhibitory compounds, such as phenols and other recalcitrant structures. Vinasse contains a rich mineral and organic composition, which could theoretically used as nutrients and support the growth of fungi of biotechnological interest. Considering the high volumetric production of vinasse, there is a significant potential of producing value-added compounds through the utilization of microorganisms that can efficiently grow in vinasse. While it has been demonstrated that the nutrients in vinasse can support the cell growth and lipid accumulation of a wild strain of the zygomycete *Mucor circinelloides* (Reis et al., 2019), the presence of phenolics compounds demonstrate to pose considerable inhibitory effects on both lipid and biomass productivities by *M. circinelloides.* In this sense, this work analyzes the effect of different homogeneous oxidation processes used to pre-treat vinasse prior to fungal culturing. Advanced oxidation processes (AOP) represent a wide array of wastewater treatment options. The principle of most AOP techniques is involved with the high reactivity of radicals generated, which promotes oxidation of all sorts of pollutants and recalcitrant components in wastewater. Multiple AOP have been described in the literature aiming to degrade molecules that present are little to no biodegradable, including those based in ultraviolet irradiation (UV), Fenton reaction, and the oxidizing power of peroxides. With this in mind, vinasse was treated using a batch system with the addition of Fe^{2+} (Fenton), in the presence of UV, and with the addition of hydrogen peroxide (H_2O_2). The experimental design followed a full factorial design, evaluating the effect of Fenton, and Fenton combined with UV and with H_2O_2, under two different pH values, i.e., 4.5 and 10. Results show that all the reaction systems were able to degrade part of the Chemical Oxygen Demand (COD), ranging from 18.4% up to 54.1% in the factorial results. All the oxidation assays were also able to degrade the phenolic compounds present in vinasse, achieving values as high as 86.2% in the system based in the combination of Fenton with H_2O_2 addition assisted by UV irradiation. The pre-treated vinasse samples were assayed for the growth performance of *M. circinelloides*, demonstration that the control value of 3.54 $g_{biomass}$ $L^{-1}_{vinasse}$ was increased to values as high as 4.84 $g_{biomass}$ $L^{-1}_{vinasse}$ when the vinasse pre-treated by Fenton assisted by peroxide addition was used as culture medium. In addition, the control conditions indicated a lipid productivity of 152 mg L^{-1} day^{-1}, which was increased by about 73% (265 mg L^{-1} day^{-1}) when the vinasse treated by UV-assisted fenton system was used. Therefore, this work demonstrates the potential of integrating efficient homogenous Fenton-based systems to the pre-treatment of vinasse in order to increase its feasibility as culture media for fungi of biotechnological interest.

Ref.: Reis, C. E. R., Carvalho, A. K. F., Bento, H. B., & de Castro, H. F. (2019). Integration of microbial biodiesel and bioethanol industries through utilization of vinasse as substrate for oleaginous fungi. *Bioresource Technology Reports*, *6*, 46-53.

574758 Co-Culture of Filamentous Feed-Grade Fungi and Microalgae As an Alternative to Increase Feeding Value of Ethanol Coproducts

Distiller's grains, an important commodity in the feed and food chains, are currently underdosed in rations due to several factors, mainly nutrient imbalance. This study aimed to increase the linoleic acid content in distiller's grains and decrease the excess nutrients in stillage water by the use of an artificial lichen, composed of fungi, algae, and a supporting matrix. A maximum concentration of 46.25% of linoleic acid in distiller's grains was achieved with a combination of Mucor indicus and Chlorella vulgaris using corn-to-ethanol whole stillage as substrate. Microbial hydrolytic enzymes during fermentation were able to decrease the solids in whole stillage. Nitrogen depletion by microalgal uptake causes lipid-formation stress to Mucor indicus cells, increasing linoleic acid production to about 49% of the total lipids, potentially decreasing costs in the animal feed. The culture supernatant can potentially be recycled as process water to the ethanol fermentation tank, and enhanced distiller's grains can replace animal-specific diets. This would reduce exogenous enzyme use and supplementation of unsaturated fatty acids from other sources.

566683 Anaerobic Co-Digestion of Swine Wastewater and Agricultural Residues for Biogas Production

ANAEROBIC CO-DIGESTION OF SWINE WASTEWATER AND AGRICULTURAL RESIDUES FOR BIOGAS PRODUCTION

Sabarish Srinivasan[1], Jay J. Cheng[2], Qingfang Zhang[3]

[1] Panther Creek High School, Cary, NC 27519, USA

[2] Department of Biological and Agricultural Engineering, North Carolina State University, Raleigh, NC 27695, USA

[3] Lanzhou University of Technology, Lanzhou, Gansu Province, China

1. INTRODUCTION

The rapid and irreversible depletion of fossil fuels accompanied by an increasing global desire for energy has led to the search for alternative energy sources. Biogas, composed mainly of methane and carbon dioxide, can provide electricity generation along with heating in place of nonrenewable energy choices. An efficient technique for producing biogas is anaerobic digestion, a microbial decomposition of organic matter that naturally occurs without oxygen. The addition of co-substrates in anaerobic digestion is known as anaerobic co-digestion: the breakdown of several organic waste varieties in the same "digestor" or bioreactor. The utilization of agricultural residues as co-substrates in the co-digestion process is especially appealing due to the reduction of agricultural waste as well as the biogas formation provided. This mechanism generally increases the amount of biogas yielded due to positive synergistic effects between microorganisms in the digestion and other factors such as optimization of the C:N ratio [1]. A wide variety of feedstocks may be utilized, but a major option is lignocellulosic biomass: plant material, composed of lignin, hemicellulose, and cellulose, that is high in carbon content, contributing to the balance of the C:N ratio for better digestion [2]. This study was designed to investigate lignocellulosic biomass as a co-substrate in order to determine the means of obtaining the highest yield in terms of biogas creation in anaerobic co-digestion.

2. MATERIALS AND METHODS

In this study, the volumetric biogas output ability of disparate substrates in the anaerobic co-digestion process was tested in four batch bioreactors operating at a temperature of 35°C in a computerized respirometer. The batch reactors were four 500 mL glass bottles that each contained two ducts with tubes attached to sustain gas collection and to sample the biogas. The gas produced was measured by 500 mL volume cylindrical liquid displacement meters over a period of 30 days. To maintain the anaerobic environment, the bioreactors were purged of oxygen through the addition of nitrogen at the beginning of each trial. Additionally, the reactors were immersed in water that reached approximately ¾ of their height to maintain their temperature at 35°C while running. All four reactors were provided with 300 milliliters of anaerobic culture and 200 milliliters of swine wastewater. The culture was derived from a 14-liter continuous stirred tank reactor undergoing anaerobic co-digestion with swine manure and switchgrass operated at mesophilic conditions at 35°C and with a hydraulic residence time of 25 days. Furthermore, 5 grams of an agricultural residue were added once initially to each reactor: switchgrass, corn stover, rice straw, or wheat straw. The amount of co-substrate in the experiment was changed after 9 days of experimentation due to unyielding, lower than expected stabilization in biogas production. In this new trial, 3 grams of corn stover, 8 grams of rice straw, and 8 grams of wheat straw, were implemented while the portion of switchgrass was not changed. For this test, the experiment was completely restarted under the same conditions as the initial trial except for the new co-substrate totals.

3. RESULTS

For the first 4 days of the experiment, the biogas production of the co-digestions of corn stover, rice straw, and wheat straw, fluctuated rapidly before leveling and maintaining a consistent output from the 4th day to the 9th day: 90 mL for corn stover, 170 mL for rice straw, and 155 mL for wheat straw. After the 9th day and the alterations to the quantities of co-substrates utilized, the corn stover biogas production presented a decline of approximately 40%. The rice straw biogas yield did not demonstrate a remarkable change with a 3% increase over the 30-day period. Conversely, the gas creation in the switchgrass co-digestion exhibited an instantaneous exponential growth trend throughout the first 9 days, and this inclination continued until the termination of the study. The output for the wheat straw, on the other hand, exhibited an increase of 150% in two days and an overall increase of 50% in the second phase of the testing, passing switchgrass biogas volume in the latter half of the trial. The wheat straw biogas production maxed at 53.75 mL/g biomass while the switchgrass biogas formation maxed at 48 mL/g biomass during the second phase of the experimentation. The results indicate that biogas production from the switchgrass is steadier and has maintained a higher conversion efficiency that that of the wheat straw over the entire trial period. Moreover, the switchgrass biogas formation surpassed that of wheat straw by a large margin of 67 mL/g, demonstrating that this residue is exceptionally effective in comparison to the other substrates used in the study.

4. DISCUSSION

The outcomes of the experiment indicate that switchgrass is the most effective co-substrate because of its high biogas production rate. The chemical composition of the dry agricultural residues reveals that the cellulose, hemicellulose, and lignin (lignocellulosic elements) proportions differ among the co-substrates. The average switchgrass cellulose percentage is 39.5% which exceeds that of the other residues, and the average switchgrass hemicellulose percentage is 20.3% which undershoots that of the other residues, suggesting a correlation between cellulose, hemicellulose, and biogas generation due to switchgrass performance [3]. One factor that could explain the wheat straw production is the relatively low lignin (17%) portion which could have accelerated the digestion process because lignin cannot be decomposed in anaerobic digestion without pretreatment. Switchgrass did display more consistent biogas growth when compared to wheat straw; however, further testing must be conducted to ascertain the possibility of altering conditions to augment the biogas released in the co-digestion of wheat straw. Additionally, the usage of switchgrass may be more efficient because the switchgrass quantity does not have to be changed for optimal results, unlike the other co-substrates. The physical attributes that should be considered in the application of lignocellulosic substrates including fiber length, fiber strength, cell wall thickness, and bulk density, can also play a significant role in their utilization in agricultural practices. For example, switchgrass is generally a sturdier substance due to these characteristics than wheat straw, and, therefore, can handle compression forces and stress with greater ease, furthering its applicability to industrial settings. Another factor that may influence the scalability of this investigation is the small quantities of co-substrates wielded, ranging from 5 grams to 8 grams. Extrapolation would be necessary for execution in energy or other relevant sectors, and further studies should be organized with larger amounts of co-substrates before leveling up to industrial application.

5. CONCLUSION

The study of switchgrass, wheat straw, rice straw, and corn stover in the anaerobic co-digestion process by batch mode reveals that switchgrass is the best co-substrate of those that were tested in this experiment. The combination of chemical and physical properties that aid in the rapid degradation of this material and conversion to biogas was unmatched in the first trial, exhibiting quick exponential biogas development. Although the wheat straw biogas production did rival that of switchgrass in the second trial, the increased quantity that was necessary for this change does not justify the marginal difference in biogas output, illustrating that, overall, switchgrass is the most efficient, reliable, and productive agricultural residue in this investigation of the anaerobic co-digestion of swine manure and certain co-substrates.

REFERENCES

1. Mata-Alvarez, J., Macé, S., & Llabrés, P. (2000). Anaerobic digestion of organic solid wastes. An

overview of research achievements and perspectives. *Bioresource Technology, 74*(1), 3-16.
2. Gontupil, J., Darwin, M., Liu, Z., Cheng, J. J., & Chen, H. C. (2012). Anaerobic Co-Digestion of Swine Manure and Corn Stover for Biogas Production. *2012 Dallas, Texas, July 29 - August 1, 2012*.
3. Kim, D. (2018, 02). Physico-Chemical Conversion of Lignocellulose: Inhibitor Effects and Detoxification Strategies: A Mini Review. *Molecules, 23*(2), 309.

574220 Effect of Fenton Pretreatment on Cellulose Oxidation and Its Enzymatic Hydrolyzability

Cellulose oxidation plays an important role in improving enzymatic hydrolysis by making cellulose more accessible. To determine whether C1 or C6 oxidation in Fenton pretreatment can boost hydrolysis rate of cellulose, we examined the structural changes of cellulose by ^{13}C NMR and correlated the C1 oxidation with the hydrolysis yield of Fenton pretreated Avicel. The sequential alkaline (NaOH) and Fenton pretreatments (AFP) significantly increased the 72 h hydrolysis yield of Avicel by 35% as compared to the control group. Langmuir adsorption isotherms showed that Fenton pretreatment increased the adsorption capacities (G_m) but decreased the affinity between cellulase enzymes and cellulose, which resulted in a higher 72 h hydrolysis yield but a lower initial hydrolysis rate. Spectra from ^{13}C NMR showed that C1 and C6 oxidations of AFP-Avicel were 19.6% and 3.5%, respectively. A good correlation was observed between the 72 h hydrolysis yield and C1 oxidation. The results indicated that the C1 oxidation generated more accessible sites for cellulase enzyme hydrolysis which led to a higher 72 h hydrolysis yield. However, observations of C6 oxidization were consistent with the reduction of enzyme binding by electrostatic repulsion and a lower initial hydrolysis rate. Cleavage of glycosidic bonds and oxidation of C1 by Fenton reaction is similar to lytic polysaccharide monooxygenases (LPMOs).

581709 Filament Fabrication and 3D Printability of Poly(3-hydroxybutyrate-co-3-hydroxyvalerate) (PHBV)/Poly(lactic acid) (PLA) blends using a Commercial Chain Extender

A series of poly(3-hydroxybutyrate-*co*-3-hydroxyvalerate) (PHBV) and poly(lactic acid) (PLA) (60:40), (50:50), and (40:60) %wt. based blends were successfully 3D printed using a commercial chain extender (CE). Breakage and buckling behaviour of the filaments during melt extrusion and 3D printing were addressed accordingly. The incorporation of the CE allowed high printing temperatures (220°C) of PHBV, resulting in good bed adhesion and warpage prevention. When investigating the 3D printing parameters of the blends, a bed temperature of 60°C and a layer height of 0.25mm were found to exhibit the highest mechanical properties. Scanning electron microscopy of tensile fractured specimens revealed that good weld formation was present for all samples that were printed at the optimal printing parameter. Poor adhesion between layers was observed when printing the blends with a bed temperature of 90°C. This study highlights the effect of the CE on the printability of PHBV/PLA based blends as well as the importance of appropriate printing parameters in the performance of 3D printed parts.

Acknowledgements: This research is financially supported by the Ontario Ministry of Agriculture, Food and Rural Affairs (OMAFRA), University of Guelph, Bioeconomy Industrial Uses Research Program Theme Project #030252; the Natural Sciences and Engineering Research Council (NSERC), Canada Discovery Grants Project #400320; and the Ontario Ministry of Economic Development, Job Creation and Trade ORF-RE09-078 (Project #053970);.

Keywords: 3D printing, poly(3-hydroxybutyrate-3-*co*-3-hydroxyvalerate), poly(lactic acid), fused deposition modeling, biopolymers.

557689 Evaluation of Binderless Board Made from Composted Rice Straw As Substrate for Rice Seedling Production

China is the largest rice producer and consumer in the world; approximately 20% of arable China's lands is annually devoted to planting this valuable crop. With the increase of China's population, the rice yield will increase to 7850kg/ha by 2030, resulting in a total of 200 million kg of rice per year to feed the Chinese people. Therefore, China has the potential to produce a large amount of rice straw. Today, compost plays the major role as a peat substitute. Composts straw have beneficial attributes, such as providing plant nutrients, suppressing disease, promoting the growth of beneficial microbes. In this paper, the straw was composted under controlled condition to form the rice seedling substrate. After the composted straw was beaten for an hour, the substrate block with dimension of 27 cm ×55 cm × 1 cm were formed by vacuum adsorption from water.

The nutrient in the substrate block, including organic matter, nitrogen, phosphorus and potassium, was improved significantly in comparation with the traditional substrate. The results of rice seedling grown in straw-based substrate block showed that the seedling height, stem diameter, root length, leaf area, SPAD of rice produced from the substrate block are nearly the same as the normal substrate. And after transplanted, the effective tillering number improved 7.4%, the yield of the rice improved 5.7%, the ratio of straw to grain decreased 11.5%. That is to say, the straw based substrate can reduce ineffective tillering, and decrease the amount of produced straw.

The substrate block can substitute the seedling tray and reduce the agricultural white pollution. Agricultural wastes in rice production can be utilized and recycled. In addition, the labor intensity and cost of rice seedling production can be reduced, and the rice production efficiency was improved.

572555 A Closed-Loop Biorefinery for Woody Biomass Conversion Using Lignin-Derived Deep Eutectic Solvents

Lignocellulosic biomass is an attractive energy source; however, conversion of the biomass is still technically and economically challenging due to its rigid and compact structure and recalcitrant nature toward conversion caused by chemical and physical features of the biomass. Therefore, effective and innovative conversion technologies are eagerly needed for converting the biomass to fuels and industrial chemicals.

Deep eutectic solvents (DES) pretreatment is an effective method for biomass conversion and/or fractionation. DESs have been proposed as alternatives to ionic liquids (ILs). These solvents do not only share the typical properties of ILs such as low volatility, wide liquid range, non-toxicity, and biocompatibility, but they also have some advantages over the traditional ILs like easy synthesis and relatively inexpensive components, thus large scale application is feasible. Tuning properties of DESs by replacing the hydrogen bond donor (HBD) with phenolics sharing structural similarities with lignin can facilitate better solubility of lignin during the fractionation. Relatively mild pretreatment conditions minimize the unwanted lignin modification and keep relatively intact structures during the pretreatment; therefore, it promotes subsequent lignin valorization.

Herein, a biomass conversion and fractionation process employing novel deep eutectic solvents (DES) prepared from lignin-derived phenolic compounds is studied. With choline chloride as HBA, the proposed process utilized the fractionated lignin-derived compounds as an HBD to enhance the total biomass utilization. Different phenolic acids such as 4-hydroxybenzoic acid and caffeic acid were investigated for effective lignin fractionation and sugar release from woody biomass (poplar). To evaluate the impact of DES pretreatment on biomass conversion, enzymatic digestibility was conducted. Diverse analytical techniques including 2D HSQC NMR, ^{31}P NMR, GPC, and wet chemistry using HPLC were performed to elucidate the structural changes during the reactions. Recyclability of DES was also studied to develop an economically feasible closed-loop system.

572703 Reuse of Sustainable Wastes for Fibreboard Production: The Case of Waste Paper and Water Hyacinth

In this early study, new cement bonded bi-composite fibreboard (FB) was made from water hyacinth (WaHy) and waste paper (WP). Ordinary Portland cement (PdCe) used as a binder was mixed with other two additives; gypsum plasters (GyPl) and wood ash (WdAh) in defined proportions to form bonding matrices. The WP and WaHy were pre-treated and a linked process developed for the mixing and consolidation steps. The FBs produced were based on different proportions of composites, binder and additives mixed. Equal amount of composites mixed with binder and additive in ratio 2:1 by weight resulted in FB having acceptable densities properties. Improved tensile strength was observed for bi-composite FB from WaHy mixed with WP. In general, FB having densities in the range of 0.50 – 0.57 gcm^{-3} compared favourably with the ASTM and ANSI standards (95%), thus making the FBs a potential alternative for building and construction purposes. More elaborate research with advanced analytical techniques are hereby suggested.

582073 Polyhydroxyalkanoate and Cellulose Ester Based Biodegradable Plastic Blends for Sustainable Packaging

Poly(3-hydroxybutyrate-co-3-hydroxyvalerate) (PHBV) has positive biodegradable, non-toxic, sustainable, and compostable properties, making it attractive for use in packaging. However, the applications in flexible packaging are limited due to its brittleness and narrow processing temperature range [1]. To mitigate the brittleness of PHBV, a compostable polymer such as cellulose acetate (CA) can be used with a 2.45 degree of substitution to positively influence the toughness of PHBV [2]. Furthermore, based on the solubility parameters, PHBV and cellulose acetate are potentially miscible [3], [4].

In this research, PHBV was blended with CA to investigate the overall mechanical properties of the blend; however, CA requires plasticization in order to be melt compounded. Prior to processing, CA powder was plasticized with triethyl citrate (TEC) and rested at room temperature for 12 hrs. Subsequently, PHBV pellets were melt compounded with the plasticized CA (pCA) powder in a twin-screw extruder followed by injection moulding. Our study found 70/30 wt.% PHBV/pCA blend ratios had improved impact strength, exceeding virgin PHBV and pCA which is attributed to PHBV degradation from the high processing temperature. TEC plasticizer migrated during processing into the PHBV, partially plasticizing it, thus reducing its melt temperature and forming secondary crystallite structures in the form of a double melt peak from differential scanning calorimetry (DSC) study. Using CA as a filler in PHBV without TEC reduced the tensile and flexural strength, and had poor interfacial adhesion from scanning electron microscopy (SEM) study. The impact strength however, was improved at a 70/30 wt.% PHBV/CA blend ratio with 50% higher tensile modulus and strength, and flexural strength compared to PHBV/pCA 70/30 wt.% blends.

Acknowledgment

This research is financially supported by the Natural Sciences and Engineering Research Council (NSERC), Canada Discovery Grants (Project # 400320); the Ontario Ministry of Agriculture, Food and Rural Affairs (OMAFRA) – University of Guelph, the Bioeconomy Industerial Uses Research Program Theme (Project # 030255); and the Ontario Ministry of Economic Development, Job Creation and Trade ORF-RE09-078 (Project #053970)

References

[1] K. Khosravi-Darani and D. Z. Bucci, "Application of Poly(hydroxyalkanoate) In Food Packaging: Improvements by Nanotechnology," *Chem. Biochem. Eng. Q.*, vol. 29, no. 2, pp. 275–285, 2015.

[2] R. M. Gardner, C. M. Buchanan, R. Komarek, D. Dorschel, C. Boggs, and A. W. White, "Compostability of cellulose acetate films," *J. Appl. Polym. Sci.*, vol. 52, no. 10, pp. 1477–1488, 1994.

[3] B. Wang, J. Chen, H. Peng, J. Gai, J. Kang, and Y. Cao, "Investigation on Changes in the Miscibility, Morphology, Rheology and Mechanical Behavior of Melt Processed Cellulose Acetate through Adding Polyethylene Glycol as a Plasticizer," *J. Macromol. Sci. Part B Phys.*, vol. 55, no. 9, pp. 894–907, 2016.

[4] J. S. Choi and W. H. Park, "Effect of biodegradable plasticizers on thermal and mechanical properties of poly(3-hydroxybutyrate)," *Polym. Test.*, vol. 23, no. 4, pp. 455–460, 2004.

582775 Extraction of Eucalyptus Globulus Leaves with Distinct Methods and Solvents. Comparison and Analysis of the Extracts

The increased awareness and research for improvements and valorization of by-products from established processes is marking the recent years as an important moment for biorefinery processes. Under the biorefinery scope, several biomasses are regaining interest for valorization, as the knowledge of sources and product certification rises. Essential oils are a classical example of a different approach for forestry residues valorization, coexisting with a different class of extracts obtained from biomass extraction, plant extracts [1].

In the present work, distinct methods were employed for *E. globulus* leaves extraction. Conventional techniques, namely Soxhlet, and solid-liquid extraction in batch mode (SLE-batch) at four different temperatures (20, 35, 55 and 75 °C), were performed with solvents of distinct polarity: dichloromethane, ethanol, methanol, *n*-hexane, petroleum ether, ethyl acetate and isopropanol. To obtain the essential oil of the *E. globulus* leaf, hydrodistillations were performed. Supercritical CO_2 extractions were also performed at 300 bar and 40 °C, with the addition of modifiers to CO_2, namely ethanol and ethyl acetate, in contents of 5-10 wt.%.

All the extracts produced, along with the ground *E. globulus* leaf, commercial samples of *E. globulus* leaf oil obtained by steam-distillation from Portugal and India, pure eucalyptol and betulinic, oleanolic and ursolic acids, were analyzed with Fourier transform infrared – attenuated total reflectance (FTIR-ATR) technique.

To avail from all the information of the FTIR-ATR spectra, a statistical tool, multidimensional scaling (MDS), was employed. MDS enables the representation of similarities/dissimilarities between complex data as distances in a low-dimensional space, allowing its visual interpretation and analysis [2].

The extractions performance varied significantly with the method and solvent employed, along with the temperature in the SLE-batch runs. The yields of extraction obtained ranged from 0.14 wt.% to 37.80 wt.% for hydrodistillation and Soxhlet extraction with methanol, respectively. All SLE-batch experiments scored lower than the Soxhlet ones. Furthermore, within the SLE-batch experiments, temperature had a clear effect, as lower temperatures yielded less and *vice-versa*. The supercritical extractions yields were similar to the less polar solvents, but higher than those of hydrodistillation.

MDS provided a scale reduction from 1000 D to 2 D with a stress factor of 9.9 %, considered good [2]. From its visualization, a clear distinction of the extraction methods employed can be observed, along with solvent polarity and temperature effect.

In this work, an experimental and modeling study on *E. globulus* leaf was performed, aiming at mapping the similarities/dissimilarities between natural products that can be produced by extraction from this biomass.On the whole, this study is expected to shed light on the biorefinery opportunities for this biomass part, including a valid enrichment of product portfolio options or market segments.

References

1. 1. G. Vânia G. Zuin, L.Z. Ramin, Green and Sustainable Separation of Natural Products from Agro-Industrial Waste: Challenges, Potentialities, and Perspectives on Emerging Approaches (Springer, 2018)
2. 2. Borg, P.J.F. Groenen, P. Mair, Applied Multidimensional Scaling (Springer, Berlin, 2013)

567224 Discrete Element Modeling of Granular Flow of Flexible Woody Biomass Particles

Biomass source materials such as pine chips, corn stover and chopped switchgrass are all irregular-shaped, deformable particles. Unlike traditional granular materials that are usually hard, biomass particles exhibit significant deformability in feeding and handling processes even under low-pressure compression. Such deformability has a profound impact on bulk flow behavior of biomass materials and makes it difficult for design of feeding and handling equipment. This work presents a bonded-sphere Discrete Element Method (DEM) approach for simulating and studying the flow behavior of irregular-shaped, deformable particles. In this approach, an individual particle is modeled by a cluster of elastically bonded spheres. The model is designed to simultaneously capture irregular particle shapes and particle-level deformations such as compression, deflection, and distortion. This approach is applied to simulate and study the behavior of a woody biomass source material of interest (i.e., fine chips) in a cyclic compression and hopper flow test. With calibration, the bulk densities and bulk moduli of elasticity calculated from the DEM simulations agree reasonably with those measured in the physical test. A sensitivity study of the impact of particle shapes and deformability on bulk flowability is conducted through simulations of a hopper test. Results show that hard particles have higher change to suffer from clogging issues, whereas deformable particles have less chance to. This indicates that particle deformation reduces the critical bridging width in hopper. Overall, this work has demonstrated the necessity to take into account of the particle deformability for the simulations of biomass particles in working conditions.

578013 Solvent and Processing Conditions for Pretreatment and Dissolution of Cellulose

Efficient utilization of biomass is hindered by the recalcitrance to dissolution of semicrystalline cellulose. [Alexandridis, P.; Ghasemi, M.; Furlani, E. P.; Tsianou, M., Solvent processing of cellulose for effective bioresource utilization. Curr. Opin. Green Sustainable Chem. 2018, 14, 40-52; DOI: 10.1016/j.cogsc.2018.05.008] Pretreatment is often used to alter the structure of cellulosic biomass in order to make cellulose more accessible to solvents and enzymes. The pretreatment involves physical and/or chemical processing which affects the degree of crystallinity and size of biomass particles. We examine here the effects of (i) solvent properties, pretreatment steps and temperature, and (ii) fiber diameter and degree of crystallinity, on the kinetics of cellulose swelling and dissolution. To this end we have combined (a) experimental results on cotton fiber swelling, change in crystallinity and dissolved amount when treated under different solvent conditions, with (b) a phenomenological model that accounts for the phenomena governing the dissolution of solid cellulose, e.g., solvent penetration, transformation from crystalline to amorphous domains, specimen swelling, and polymer chain untangling. [Ghasemi, M.; Singapati, A. Y.; Tsianou, M.; Alexandridis, P., Dissolution of semicrystalline polymer fibers: Numerical modeling and parametric analysis. AIChE Journal 2017, 63 (4), 1368-1383. DOI: 10.1002/aic.15615] [Ghasemi, M.; Alexandridis, P.; Tsianou, M. Cellulose dissolution: Insights on the contributions of solvent-induced decrystallization and chain disentanglement. Cellulose 2017, 24 (2), 571-590; DOI: 10.1007/s10570-016-1145-1] The insights obtained from this analysis would facilitate the rational selection of solvents and the design of pretreatment processes to reduce the size and degree of crystallinity of cellulosic biomass particles, leading to enhanced biomass utilization. [Ghasemi, M.; Alexandridis, P.; Tsianou, M., Dissolution of cellulosic fibers: effect of crystallinity and fiber diameter. Biomacromolecules 2018, 19 (2), 640-651; DOI: 10.1021/acs.biomac.7b01745] [Ghasemi, M.; Tsianou, M.; Alexandridis, P., Assessment of solvents for cellulose dissolution. Bioresource Technol. 2017, 228, 330-338; DOI: 10.1016/j.biortech.2016.12.049] [Ghasemi, M.; Tsianou, M.; Alexandridis, P. Population ensemble modeling of biomass dissolution. Chem. Eng. J. 2018, 350, 37-48; DOI: 10.1016/j.cej.2018.05.167]

574515 Effectiveness of Cotreatment during Fermentation on Various Feedstocks, Impact on Different Microbes, and Milling Energy Requirements

Milling during fermentation, also known as cotreatment, is being explored as an alternative to thermochemical pretreatment for increasing the accessibility of lignocellulose to biological attack. The feasibility of cotreatment hinges on achieving high carbohydrate-to-sugar conversion (also known as solubilization), the fermentative microorganism's tolerance to milling, and the energy requirements of milling. This talk presents results on different aspects of investigating cotreatment's feasibility; specifically, its effectiveness on different feedstocks, impact on various microorganisms, and milling energy requirements. Cotreatment in combination with CBP fermentation by *Clostridium thermocellum* at low solids concentrations was performed using various feedstocks: switchgrass (with and without genetic modification aimed at decreasing recalcitrance), more and less recalcitrant natural variants of *Populus*, and corn stover. The ability to carry out fermentation in the presence of aggressive milling is tested using the following microorganisms: *Clostridium thermocellum, Thermoanaerobacterium saccharolyticum, Escherichia coli*, *Bacillus subtilis,* and *Zymomonas mobilis.* As a first step towards determining energy requirements of cotreatment, substantial changes in the viscosity of corn stover slurries over the course of *Clostridium thermocellum* fermentation were evaluated at solids loadings representative of an industrial process. Moreover, variations of ball and disc milling have been screened for solubilization effectiveness. Ongoing work is focused on measuring the milling energy of fermented biomass; initial results will be presented on this subtopic.

568850 Influence of Industrial Fractionation Methods on Lignin Structure

As the only large-scale source of natural aromatics, the potential value of lignin as an alternative feedstock for petroleum-derived chemicals has sparked interest in recent years. A major hurdle for all pathways for chemically valorizing lignin is the recalcitrance of the lignin structure due to chemical transformations during fractionation. The process of fractionating lignin from cellulose and hemicellulose creates reactive sites in the lignin structure that condense to form very stable C-C bonds. The prevailing strategy to isolate lignin from biomass is Kraft pulping that generates black liquor, containing lignin polymers that show diminished promise for upgrading to chemicals. As alternatives to Kraft lignin, many new biorefineries isolate lignin using milder solvents and reaction conditions, but many of these lignin structures have yet to be thoroughly characterized. Broad characterization of lignin isolated using industrially feasible methods will aid in developing tailored valorization strategies.

Herein, eight widely used analytical techniques were used to characterize the structures of eight industrial lignin samples from wood to determine the effects different industrial fractionation and extraction methods have on the purified lignin stream. The scope of characterization covers hardwood and softwood lignin extracted using Kraft, super-critical water, and organosolv fractionation processes. Molecular weights were determined through gel permeation chromatography (GPC). Nuclear magnetic resonance (NMR) spectroscopy and infrared spectroscopy were used to identify common linkages, prevalence of various monolignol units, and functional moieties. Thermogravimetric analysis (TGA) provided insight to the thermal-physio properties of the lignin polymer. Extensive characterization of the lignin allows for correlations between the fractionation and extraction methods and final lignin structure, as well as, a comparison between the utility of the many characterization techniques. The non-Kraft lignin samples, in general, show lower levels of recalcitrance compared to the Kraft lignin, but still exhibit significant alteration to the original structures.

575103 Molecular Reconstruction of Esterified Pyrolysis Bio-Oil Using Fourier Transform Ion Cyclotron Resonance Mass Spectrometry

In this work was modified and enhanced the methodology of structural representation through structural attributes employed in the characterization of complex mixtures as heavy petroleum oils from unconventional and conventional analytics[1]. A Colombian esterified pyrolysis bio-oil obtained by fast pyrolysis of a mixture of softwoods was characterized by molecular representation. Information obtained by Fourier Transform ion cyclotron resonance mass spectrometry FT-ICR-MS (standard procedure in house) and results of bulk properties as analysis elemental (ASTM D4052), density measure by means of a digital densimeter (ASTM D4052) and, carbon residue by micro method (ASTM D4530), were used to represent molecularly the sample. The molecules were performed by Monte Carlo technique from chemical structural attributes. The new structural attributes considered in the methodology of molecular representation corresponding to oxygen groups (-O, -OH Aromatic, -OH Aliphatic, -O- Aromatic, -O-Aliphatic, =O). Each attribute in the sample can be represented by a probability distribution function (PDF), which are optimized to adjust the structures of pyrolysis oil and its composition to the experimental data. As a result, 300 molecules were generated. The bulk properties for esterified pyrolysis oil were in good agreement with the experimental structural information.

Acknowledgments

This work was supported by a Newton Fund award (reference number 275910721), Research Agreement No. 5211770 UIS-ICP.

References

[1] C. X. Ramírez *et al.*, "Molecular Representation of Petroleum Residues Using Fourier Transform Ion Cyclotron Resonance Mass Spectrometry and Conventional Analysis," *Energy and Fuels*, vol. 31, no. 12, pp. 13353–13363, 2017.

579326 Catalytic Hydrothermal Treatment (CHT) of Lignocellulosic Biomass

Value added co-products such as bioethanol and levulinic acid can be obtained from fermentation and catalytic hydrothermal treatment (CHT) of lignocellulosic biomass (LCB). However, the commercial value of bioethanol production is significantly lower due to the challenges in biomass pretreatment, enzymatic hydrolysis, and fermentation process. Hence to accommodate the return of investment, levulinic acid production can be more beneficial. In this study, lignin and hemicellulose were separated from hydrothermal pretreatment of corn stover at 170-190°C with 0.45% w/v K_2CO_3 in water. Following the pretreatment step, the cellulose rich biomass was converted into levulinic acid using CHT process at reaction temperatures of 150- 250°C. The other by-products obtained from this CHT process were catalytically treated at 800-1000°C to obtain lignin, which could be used as electrode material for energy storage applications such as batteries and supercapacitors. The liquid hydrolysate obtained from the pretreatment process was found to be rich in C5 sugars and could be further treated via wet-oxidation. The product streams were characterized using HPLC, GC-MS and TOC. Details of the product yields and their characterization will be presented.

573643 Synthesis and Characterization of Water-Soluble Anionic Carboxylic Acid-Terminated Lignin Dendrimer for Metal Chelating

The preparation of high-value and functional materials from renewable and low-cost biowaste (e.g. lignin) has been attracted a lot of interest. Lignin is an aromatic polymer that has usually been reclaimed from biorefinery waste stream. In this study, we aim to synthesize a water-soluble anionic dendrimer starting from lignin, and explore its capabilities for copper chelating. It is known that water-soluble anionic dendrimer has great potential as mimics of anionic micelles or proteins, EDTA similarities, and chelating agents for a spectrum of applications in catalyst, drug delivery, gene transfer, wastewater treatment, and micronutrients. Herein, a water-soluble carboxylic acid-terminated lignin dendrimer structure (LDCA) was successfully synthesized from lignin dendrimer via butyl ester termination approach. At first, the water-soluble lignin-based dendrimeric structure was firstly prepared from the commercial sodium lignosulfonate (LS) through grafting with the epoxy group which was reacted with amines. Then carboxylic acid groups were added as the end group prepared from neutralized lignin-NH_3^+ and t-butyl acrylate. The ^{1}H-NMR and FTIR characterization confirms the successful grafting of functional groups and the synthesis of lignin-based dendrimer and LDCA. Furthermore, the absorption of Cu (II) ion on the anionic LDCA complex was under investigation and the results from the FTIR and SEM demonstrate the binding of Cu^{2+}on the complex. The adsorption batch experiment was measured through an ultrafiltration method in the pH range of 3-9 as well as several ratio of metal ions and LDCA. The binding equilibrium was also analyzed by potentiometric titration combined with Isothermal titration calorimetry (ITC) which could fit a complexation binding model.

576449 Functionalized Superparamagnetic Lignin Nanosphere for Phosphate Removal

Phosphorus is an essential element for plant growth and crop production. However, excessive human activities introduce extra phosphorus into water bodies which leads to environmental degradation, such as eutrophication. Thus, it is necessary to develop strategies to remove phosphate from wastewater before it discharges into natural water bodies through runoff. However, it is big challenge to develop environmentally benigh and cost-effective absorbent for effective phosphate removal. Herein, we developed a lignin-based and recyclable superparamagnetic nanosphere to effectively remove phosphate in water. Lignin, a highly branched aromatic biopolymer in nature, has attracted many attentions since it is naturally abundant, renewable and low in cost. In this study, the lignin was firstly modified with a polyethyleneimine (PEI) together and then a PEI functionalized superparamagnetic Fe_3O_4 lignin nanosphere (LMS) was developed through in-situ coprecipitation method. The structural and morphology of the obtained LMS were characterized by TEM, SEM, VSM and FTIR. In addition, the adsorption capacities of LMS for phosphate from aqueous solutions and the effects of different conditions were also comparatively investigated. The adsorption capacity results showed that the LMS exhibited an excellent adsorption performance around 41.84 mg/g. Overall, this new developed magnetic lignin nanosphere has shown a great potential as a low cost, high adsorptive, and reusable adsorbent for the phosphorus removal.

576470 Using Soybean Hull in Production of Paper-Based Gas Sensor for Food Packaging Applications

Using soybean hull in production of paper-based gas sensor for food packaging applications

Soybean hull is a major co-product of soy industry. It is the seed coat of the soybean that is removed during the grinding process and represents eight percent of the total weight of the biomass. Global annual production of soybean hull is 18-20 million tons with approximately 10 million tons being generated in USA in 2018. Some of the current applications of this co-product include utilization of it as animal feed, in wastewater treatment, as dietary fiber, and in herbal medicine. With the objective of soybean hull valorization towards production of high value products, we have recently utilized processed soybean hull as a novel binder in the fabrication of paper-based gas sensor for food safety. The production of this novel binder requires only traditional, simple to apply, processing steps and is environmentally friendly. The paper-based sensor was fabricated by coating bentonite, soybean hull-based binder and a pH sensitive dye (bromocresol purple) on paper in three different coat weights. Use of soybean hull product as a binder decreases gas sensor production costs drastically comparing to any other commercial binder, including polyvinyl alcohol (PVOH) which has been used in our previous study. Furthermore, application of the soybean hull product can enhance the sensor's gas capture ability by providing high void volume. Lower binder demand of bentonite makes the process even more cost-efficient. Bentonite adsorbs volatile basic nitrogen compounds (VBN) which are metabolic by-products of spoilage bacteria. The prepared samples of gas sensor were tested for their ability to detect VBN through distinct color change from yellow to purple. For this purpose, ammonium hydroxide solution with different concentrations were used. The highest sensitivity and durability were 25 μL (upon exposure to 1mM ammonium hydroxide solution) and 24 hr (upon exposure to 100 μL of 5mM ammonium hydroxide solution), respectively (Figures 1 and 2). These results are comparable to the ones obtained for the PVOH-containing sensors.

Keyword: Gas sensor, Food safety, Soybean hull

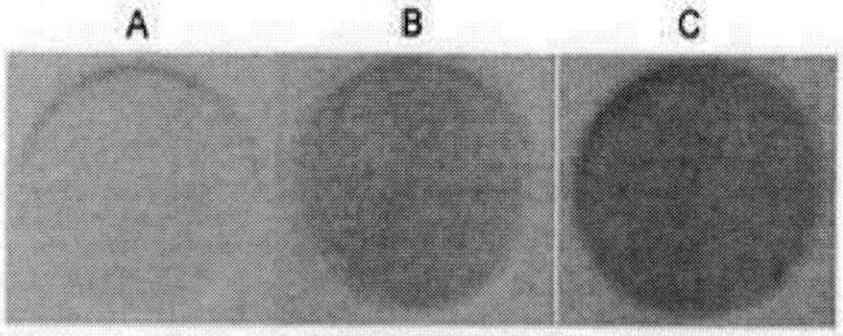

Figure 1. A, B, and C are sensitivity images upon exposure to 25, 50, and 100 μL of 1mM ammonium hydroxide solution, respectively

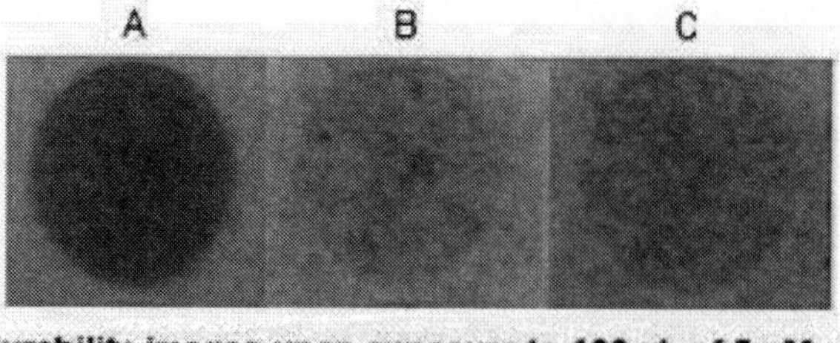

Figure 2. A, B, and C are durability images upon exposure to 100 μL of 5mM ammonium hydroxide solution, after 12 hr, and after 24 hr, respectively

557564 The Application of Nonlinear Dynamic Optimization on the Three Pseudo-Components Kinetic Modeling of Biomass (Rhus Typhina) Pyrolysis Using the DAEM

Rhus Typhina originated in North America is generally used for beverages and medicines. It contains valuable contents such as gallic acid, anthocyanins, and oleic and linoleic acids, which shows its great economic values. Rhus Typhina also has potentials as a good-quality biofuel, due to its high content of volatile matter, high HHV, and low content of minor elements such as sulfur. Despite the great potentials, the studies on the pyrolysis process of Rhus Typhina were not found in publications. In this work, the pyrolysis process of Rhus Typhina was investigated. The sample of Rhus Typhina was analyzed for the proximate and ultimate analysis to study its characteristics. A kinetic model using the distributed activation energy model (DAEM) was developed to describe the thermal decomposition of Rhus Typhina in the thermogravimetric analysis (TGA). In the DAEM, three pseudo-components were proposed to represent three major components: hemicellulose, cellulose, and lignin & other minor contents. A nonlinear dynamic optimization model was developed and integrated with the three pseudo-components DAEM. The contents of hemicellulose, cellulose, and lignin & others, the pre-exponential factors, and activation energy were set as control parameters. The optimal values of these parameters were achieved from the dynamic optimization model, based on the TGA data at various heating rates (10, 20, and 30 K/min). The role of the distribution function on the kinetic models was also examined in 7-case studies. It was found that the standard deviations of the activation energy distribution functions for cellulose and lignin had a significant impact on the kinetic modeling of Rhus Typhina.

552771 Functional and Chemical Properties of Bio-Oil Produced from Biomass Residues

Functional and Chemical Properties of Bio-oil produced from Biomass

Residues

Mudasir Akbar Shah[a], N.S.Khan[a], Vimal Kumar[b]

[a]Department of Chemical Engineering, National Institute of Technology Srinagar, J & K-190006, INDIA

[b]Department of Chemical Engineering, Indian Institute of Technology Roorkee, Uttarakhand-247667, INDIA

Abstract

In the present work the pyrolysis of different types of walnut shells (WS) has been studied at different temperatures (450 °C and 550 °C) and particle sizes (2mm and 0.1mm) at a constant heating (20 °C/min) and sweep gas flow (100 cm^3/min) rates. The different walnut shells considered are paper (PSW), thin (TSW), medium (MSW) and hard (HSW) walnut shell. Experiment has The yield for bio-oil obtained from PSW, TSW, MSW and HSW are 48.2% (550 °C) and 32% (450 °C), 41.7% (550 °C) and 36.15% (450 °C), 39% (550 °C) and 32% (450 °C), and 49.7% (550 °C) and 36% (450 °C), respectively. Maximum oil (49.7 wt%) and gas (37%) yields have been obtained for HSW at a pyrolysis temperature of 550 °C, while minimum yield is obtained for PSW and MSW at an operating temperature of 450 °C. Calorific values of bio-oil obtained from different walnut shells are 26.8, 26, 26.5 and 27 MJ/kg for PSW, TSW, MSW and HSW, respectively. Chemical composition analysis of both fractions of bio-oils (organic and aqueous) has been investigated using GC-MS, FTIR, ^{1}H-NMR and ^{13}C-NMR. It is found that the pyrolytic bio-oil composition is predominated by oxygenated compounds. The oxygen compound is imitated by the presence of oxygenated fractions such as carbonyl and carboxyl groups formed by pyrolysis of the cellulose and phenol and methoxy groups formed by pyrolysis of lignin. From the structural analysis of the bio-oil, it can be concluded that the bio-oil obtained from WS may be used as a feedstock for furnaces, as a fuel, and also used as a lubricant after small modifications.

Keywords:Thermal Cracking; Pyrolysis; Biomass; Walnut Shell; Bio-oil.

556707 Efficient Method for Synthesis of High-Value 5-Methylfurfural from 6-Deoxy Sugar in Biphasic System

5-Hydroxymethylfurfural (HMF), an versatile platform chemical derived from carbohydrates is always limited by its chemical instability. However, more stable 5-methylfurfural (MF) with similar structure has the potential to replace HMF in many applications. In this work, the method of highly efficient conversion of 6-deoxy sugar to MF catalyzed by various catalysts was proposed in biphasic system. For enhancing MF yield, the effect of catalyst species, temperature (120-170 °C), organic/aqueous mass ratio (1-5), and reaction temperature on the conversion process of 6-deoxy sugar to MF was systematically investigated. Under optimal conditions, high yield (79 %) of MF was achieved in biphasic system due to MF insolubility favorable for avoiding the MF degradation and condensation. MF with excellent chemical stability has the potential to be a bio-derived platform chemical in green industry.

556631 Reductive Fractionation: Date PALM Biomass to High VALUE Chemicals Using Commercial NiAl Alloy Catalyst

REDUCTIVE FRACTIONATION; DATE PALM BIOMASS TO HIGH VALUE CHEMICALS USING COMMERCIAL NiAl ALLOY CATALYST

Emmanuel Galiwango, PhD (student), UAE University, UAE
Ali H. Al-Marzouqi, PhD, UAE University, UAE
Abbas Ahmed Khalil, PhD, UAE University, UAE
Mahdi M. Abu-Omar, PhD, University of California, USA

Abstract

Problem Statement: UAE has over 40 million date palm trees. On average approximately a total of 40 kg waste per tree is produced and therefore huge amount of waste is generated (about 1.2 million tons annually). Most of the waste from palm trees is currently turned into compost or burned to generate heat despite the high content of cellulose, hemicelluloses and lignin. Literature has reported possibility of converting different biomass into various High Value Chemicals (HVC) considering lignocellulose components as a base feed stock. However, despite the abundancy, little effort has been registered for conversion of date palm biomass to HVCs especially from Oil rich countries. The purpose of this study is to use Reductive Catalytic Fraction to convert UAE date palm biomass to HVCs using a high-pressure high temperature Parr reactor, with an aim of providing enough knowledge of lignocellulose and optimization conditions for an overall goal of production of HVC from the same biomass and investigate the gases evolved for possible future LPG replacement from biorefinery. Methods & Experimental procedure: Thermogravimetric analysis, SEM measurements, high-pressure high temperature Parr reactor for biomass conversion in this study (see Figure 1.) was carried out; Q500 TA Instrument using N_2, 20 mL/min. 10, 15 20 and 25°C/min, controlled from 25 to 900 °C, A JEOL/EO Scanning Electron Microscope and ethanol-benzene free extractive sample were used with commercial NiAl alloy catalyst. Results: From TGA, the initial weight was recorded continuously as a function of temperature and time. Sample's Thermal stability, degradation regimes and temperature were obtained. Existence of three major lignocellulose fractions was shown and hence guarantee for anticipated HVCs from this biomass. SEM images revealed different morphologies showing minimal visible surfaces changes of the catalyst, hence indicating ability to be recycled. The process revealed significant hydrocarbon production in the gas stream and appreciable conversions in the liquid product stream. Conclusion & Significance: Applied technique can successfully depolymerize lignin complex structure to HVCs with appreciable conversions and useful gas products. The catalyst used can easily be separated out of biomass after reaction making it cheaper and easy for re-use. Recommendations are made for further characterization for better understanding of palm biomass prior to processing in a bio-refinery and further analysis of liquid product is underway to determine the selectivity and yield of individual HVCs produced.

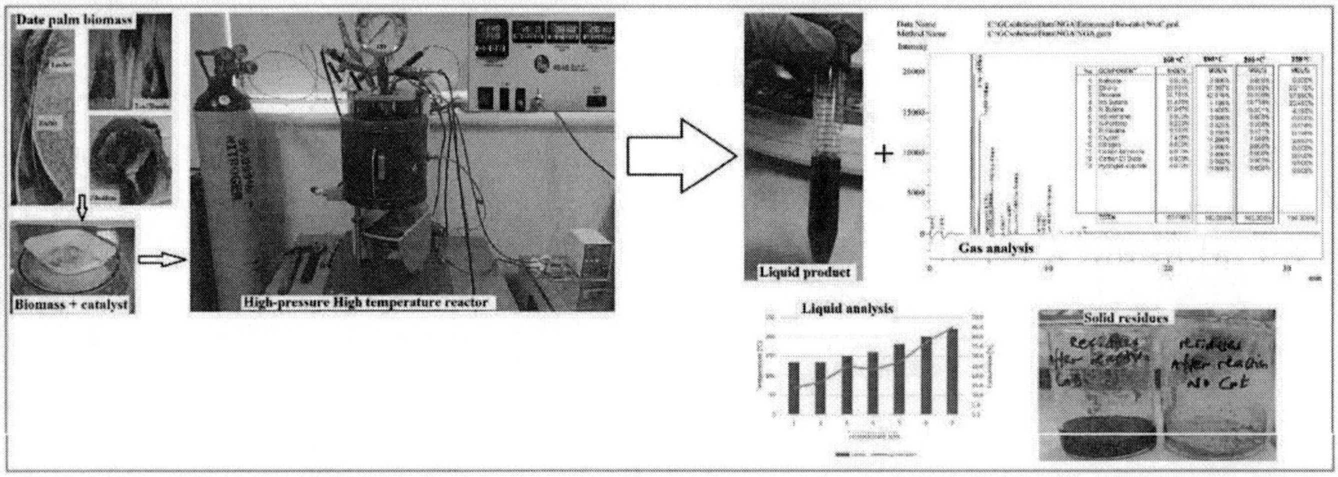

Fig. 1 Schematic layout for date palm waste to HVCs

Recent Publications

1. **Galiwango E.**, Ali H. Al-Marzouqi, Mahdi M. Abu-Omar, Abbas Ahmed Khaleel, Nour S. Abdel Rahman.Chemical and Process Engineering Research www.iiste.org. ISSN 2224-7467 (Paper) ISSN 2225-0913 (Online)Vol.57, 2018

2. **Galiwango E**., Ali H. Al-Marzouqi, Mahdi M. Abu-Omar, Abbas Ahmed Khaleel, Nour S. Abdel Rahman." Estimating Combustion Kinetics of UAE Date Palm tree Biomass using Thermogravimetric Analysis". Journal of Natural Sciences Research, 7(12), 106-120 (2017).

3. Ly H.V., **Galiwango E.**, Kim S., Kim J., Choi J.H., Woo H.C., Othman, M. R. (2017). Hydrodeoxygenation of 2-furyl methyl ketone as a model compound of algal saccharina japonica bio-oil using iron phosphide catalyst. Chemical Engineering Journal, 317, 302-308.

4. Ly H.V., Im K., Go Y., **Galiwango E.,** Kim S., Kim J. Woo H.C. (2016). Spray pyrolysis synthesis of γ-Al_2O_3 supported metal and metal phosphide catalysts and their activity in the hydrodeoxygenation of a bio-oil model compound. Energy Conversion and Management, 127, 545-553.
5. Al-Zuhair S., **Al-Marzouqi A**., Hashim I., Taher H., "Biodiesel Production", US Patent 2014/0075828 A1.
6. Taher H., Al-Zuhair S., **Al-Marzouqi, A.**, Haik Y., Farid M. (2015). Growth of microalgae using CO_2 enriched air for biodiesel production in supercritical CO_2. Renewable Energy, 82, 61-70.

Biography

Emmanuel has masters in Energy and Chemical Engineering from South Korea and he worked as a Petroleum chemist with Saybolt international before joining UAE University, where he serves as a research assistant and a PhD candidate in Chemical Engineering under Prof. Ali Al-Marzouqi's supervision. Prof. Ali Al-Marzouqi obtained his B.Sc. in Chemical Engineering from University of Washington, USA and Ph.D. in Chemical Engineering from Oregon State University, USA. He worked as an Instructor in the Chemical Engineering Department of Oregon State University for three years and then joined Chemical and Petroleum Engineering Department of UAE University as an Assistant Professor, Associate Professor; he was promoted to Full Professor and now serving as the Vice Dean for Research and Graduate Studies. Prof. Al-Marzouqi's research focuses on biomass utilization and the use of supercritical fluid technology in numerous applications including food and pharmaceutical processing, herbal extraction, biodiesel production, extraction and oxidation of oil sludge, and development of nanocatalysts.

Email:201690119@uaeu.ac.ae or emmanuelgaliwango@gmail.com

ACS Sustainable Chemistry & Engineering

Research Article

Cite This: ACS Sustainable Chem. Eng. 2019, 7, 2634−2647 pubs.acs.org/journal/ascecg

Codesign of Combinatorial Organosolv Pretreatment (COP) and Lignin Nanoparticles (LNPs) in Biorefineries

Zhi-Hua Liu,[†,‡] Naijia Hao,[§] Somnath Shinde,[§] Michelle L. Olson,[∥] Samarthya Bhagia,[§] John R. Dunlap,[⊥] Katy C. Kao,[∥] Xiaofeng Kang,[#,◇] Arthur J. Ragauskas,[§,∇] and Joshua S. Yuan*[,†,‡,∥]

[†]Synthetic and Systems Biology Innovation Hub (SSBiH) and [‡]Department of Plant Pathology and Microbiology, Texas A&M University, Norman E. Borlaug Center, 2123 TAMU, College Station, Texas 77843, United States

[§]Department of Chemical & Biomolecular Engineering, University of Tennessee, Nathan W. Dougherty Engineering Building, 1512 Middle Dr., Knoxville, Tennessee 37996, United States

[∥]Department of Chemical Engineering, Texas A&M University, Jack E. Brown Engineering Building, 3122 TAMU, College Station, Texas 77843, United States

[⊥]Advanced Microscopy and Imaging Center, University of Tennessee, Science and Engineering Research Facility, 1414 Circle Drive, Knoxville, Tennessee 37996, United States

[#]Department of Molecular & Cellular Medicine, Texas A&M University, Reynolds Medical Building, 206 Olsen Boulevard, College Station, Texas 77843, United States

[◇]Key Laboratory of Synthetic and Natural Functional Molecular Chemistry, College of Chemistry & Materials Science, Northwest University, Xi'an 710069, People's Republic of China

[∇]Biosciences Division, Oak Ridge National Laboratory, Oak Ridge, Tennessee 37830, United States

Supporting Information

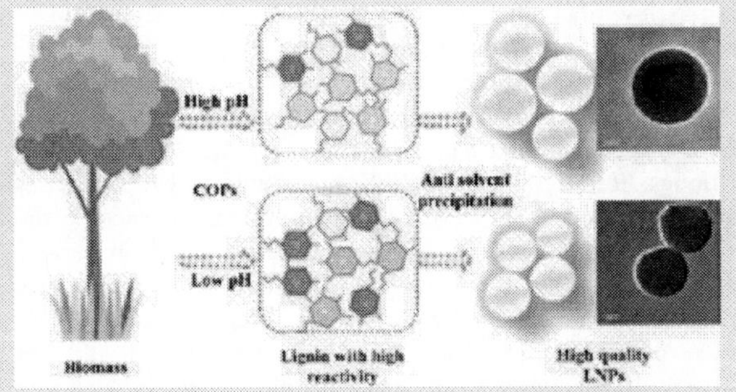

ABSTRACT: To make biorefineries sustainable, codesign of fractionation technologies and lignin valorization has been found to be essential. Combinatorial organosolv pretreatment (COP) was thus developed in an effort to efficiently produce sugars and improve lignin processability for the fabrication of lignin nanoparticles (LNPs). COP produced greater than a 90% glucose yield and 73% xylose yield, suggesting the improved sugar release from biomass. LNPs were fabricated from the lignin fractionated by COP via antisolvent precipitation. The smallest effective diameter (142 nm) of LNPs was obtained from COP using EtOH plus sulfuric acid. These LNPs possessed a lower polydispersity index and higher zeta potential, suggesting superior uniformity and greater stability. The lignin characterization results indicated that COP using EtOH plus sulfuric acid cleaved more *β*-O-4 and *β*−*β* linkages and produced lignin with a higher molecular weight and increased G-lignin and C5-substituted OH contents, suggesting the generation of condensed lignin. These modifications enhanced the hydrophobic interactions between lignins and thus enabled the fabrication of LNPs with a small particle size. COP using EtOH plus sulfuric acid also enriched total phenolic OH content and could promote the formation of a hydrogen-bonding network within LNPs. Together with a high zeta potential due to the increased phenolic OH and COOH groups, the stability of LNPs was thus enhanced. Overall, COP increased the sugar release from biomass and improved the lignin processability to facilitate the design of LNPs with satisfactory properties, which showed the potential to improve the lignin valorization and the sustainability of biorefineries.

KEYWORDS: *Combinatorial organosolv pretreatment, Lignin valorization, Codesign, Lignin processability, Lignin nanoparticles, Sugar platform*

■ INTRODUCTION

The development of competitive biorefineries has become increasingly urgent in the face of current global problems such as shortages of fossil energy and global warming.[1−4] Ideal modern biorefineries could adapt a multistream integrated concept in which complete lignocellulosic biomass (LCB) is utilized and various bioproducts are produced to achieve sustainability and feasibility.[5,6] However, the commercial implementation of biorefineries remains economically infeasible due to the lack of effective fractionation and high-value

Received: November 4, 2018
Revised: December 17, 2018
Published: December 28, 2018

ACS Publications

Table 1. Combinatorial Organosolv Pretreatment (COP) of Corn Stover and Switchgrass with Controlled pH Values[a]

		step 1					step 2				
case	feedstock	chemicals	conditions	SL	initial pH	$\log R_0''$	chemicals	conditions	SL	initial pH	$\log R_0''$
1	corn stover	liquid hot water	120 °C, 30 min	10%	6.2	2.8	50% EtOH + 1% sulfuric acid	120 °C, 60 min	10%	1.8	7.6
2	corn stover	liquid hot water	120 °C, 30 min	10%	6.2	2.8	50% EtOH + 2% formic acid	120 °C, 60 min	10%	3.8	5.6
3	corn stover	liquid hot water	120 °C, 30 min	10%	6.2	2.8	50% EtOH	120 °C, 60 min	10%	6.0	3.4
4	corn stover	liquid hot water	120 °C, 30 min	10%	6.2	2.8	50% EtOH + 1% sodium hydroxide	120 °C, 60 min	10%	13.0	8.4
5	corn stover	liquid hot water	120 °C, 30 min	10%	6.2	2.8	50% MeOH + 1% sulfuric acid	120 °C, 60 min	10%	1.6	7.8
6	switchgrass	liquid hot water	120 °C, 30 min	10%	6.3	2.8	50% EtOH + 1% sulfuric acid	120 °C, 60 min	10%	1.8	7.6

[a]SL represents solid loading; EtOH represents ethanol; MeOH represents methanol; EtOH+SA represents ethanol plus sulfuric acid; EtOH+FA represents ethanol plus formic acid; EtOH+SH represents ethanol plus sodium hydroxide; MeOH+SA represents methanol plus sulfuric acid. $\log R_0''$ is calculated based on eq 2. Corn stover was used in cases 1–5, and switchgrass was used in case 6.

byproducts. To achieve a sustainable biorefinery, conversion processes as well as lignin-derived products need to be codesigned to improve sugar fractionation and lignin valorization simultaneously.

The polymeric carbohydrates and lignin in LCB are not easily released due to the recalcitrance of the plant cell wall.[3] Various pretreatments have been employed to overcome the recalcitrance by deconstructing lignin–carbohydrate complexes and increase the accessibility of cellulose to enzymatic attack.[7−9] Pretreatment can be roughly classified into three clusters by pH value, namely, acidic, neutral, and alkaline pretreatment. In acidic pretreatment, such as with dilute sulfuric acid, most of the hemicelluloses are removed from the solid and recovered as dissolved monosaccharides in liquid, and the glucose yield from cellulose consequently increases. Although little lignin is dissolved, it disrupts and recondenses lignin in a solid.[4,8] However, acidic pretreatment leads to the irreversible degradation of hemicelluloses and generates inhibitors.[8,10,11] It may generate "pseudolignin" under very severe conditions, reducing the enzymatic conversion.[12,13] Pretreatment performed at neutral pH, such as in liquid hot water, only uses pressure to maintain pure water in its liquid state to deconstruct LCB by an autohydrolysis effect.[14] However, this process usually leads to incomplete LCB matrix opening and low hydrolysis efficiency. Alkaline pretreatment is comparatively more effective for lignin solubilization.[15] Unfortunately, possible loss of sugars and production of inhibitors may occur. In addition to these pretreatments, organosolv pretreatment is also a promising approach and more feasible for biorefinery applications.[16,17] Organosolv pretreatment solubilizes part of hemicellulose, alters the lignin structure, and removes most of the lignin. Most importantly, cellulose and lignin can be isolated as a solid material and liquor, facilitating the utilization of these components in the biorefinery.

Pretreatment can be conducted at a high holding temperature with a short residence time. This pretreatment configuration improves the hydrolysis efficiency but inevitably causes sugar losses, generates inhibitors, and requires high energy consumption due to the high temperature employed.[8,18,19] In contrast, pretreatment conducted at a low holding temperature with a long residence time requires lower pressure and heating requirements for the reactor and thus tends to avoid the serious degradation of components in LCB. Taken together, it is unlikely that a single pretreatment can be suitable for simultaneously producing high sugar yield and more reactive lignin for various products. The pretreatment option should be a compromise between positive and negative effects due to the complex effects of parameters and considerations. As new applications in the field of LCB valorization expand, pretreatment exploitation will aid viable implementation.

Lignin valorization to value-added products, such as lignin nanoparticles (LNPs), has recently been considered as an essential process for a sustainable biorefinery.[1,20−22] Because of its biodegradability, biocompatibility, and low toxicity, lignin has been considered as a new potential feedstock for fabricating environmentally friendly nanoparticles. LNPs could be potentially applied to drug delivery, wound healing, nanoglue, energy storage, environmental remediation, biocidal active substances, crop additives, and coating formulations, etc.[23−26] However, lignin valorization to LNPs remains a technological challenge given the low reactivity of native lignin. Lignin is a highly complex biopolymer synthesized from *p*-hydroxyphenylpropane (H), guaiacylpropane (G), and syringylpropane (S) via different ether and C–C bonds, such as *β*-O-4-aryl ether, *β*–*β*, *β*-5, and *α*-O-4-aryl ether linkages. Lignin units also contain various functional groups, including phenolic hydroxyl, aliphatic hydroxyl, carbonyl groups, and benzyl alcohol. These chemistries determine the reactivity, fractionation performance, and hence widespread utilization of lignin. The fractionation can modify and alter the linkages and functional groups of lignin and will define its processability for the preparation of LNPs or other products.[27−29] Richter et al. reported that Kraft and organosolv lignin can be used for LNP preparation, and the properties of LNPs are governed by the type of lignin used.[26] Qian et al. prepared uniform colloidal spheres via self-assembly from alkali lignin after acetylated modification to improve its hydrophobicity.[30] Lievonen et al. obtained spherical LNPs by using softwood Kraft lignin and found that dissolving properties of lignin depend on the hydroxyl groups of lignin and solvent used, which affect the morphology of LNPs.[25] Tian et al. extracted technical lignins by deep eutectic solvent and ethanol–organosolv to prepare two uniform LNPs, serving as excellent candidates for producing multifunctional polymer nanocomposites.[31] Despite these progresses, how pretreatment impacts on lignin chemistry and regulates its reactivity and how these tuned chemistries, such as unit type and functional groups, determine the formation and property of LNPs have not been clarified systematically. Therefore, designing a new pretreatment strategy is necessary to produce more reactive lignin suited for high-quality LNP design. Taken together, a promising pretreatment option must take into consideration its impacts on both sugar yield and lignin chemistry to generate coproducts in a biorefinery concept.

To address these challenges, combinatorial organosolv pretreatment (COP) was designed not only to facilitate the production of sugars but also enhance the processability and utilization efficiency of lignin for LNPs. Several COPs with controlled pH were evaluated to deconstruct corn stover/switchgrass, improve sugar release, and enhance lignin processability. Fermentable sugars were produced by the enzymatic hydrolysis to assess the pretreatment efficiency. In the biorefinery concept, the lignin fractionated from each COP was upgraded into LNPs via antisolvent precipitation. The morphology, size distribution, charge, and stability of the LNPs were characterized to evaluate pretreatment performance. Lignin chemistry was assessed by measuring the lignin unit type, group content, and molecular weight. These analyses should help us obtain a deep understanding of the codesign of fractionation and lignin-based products to make a sustainable biorefinery.

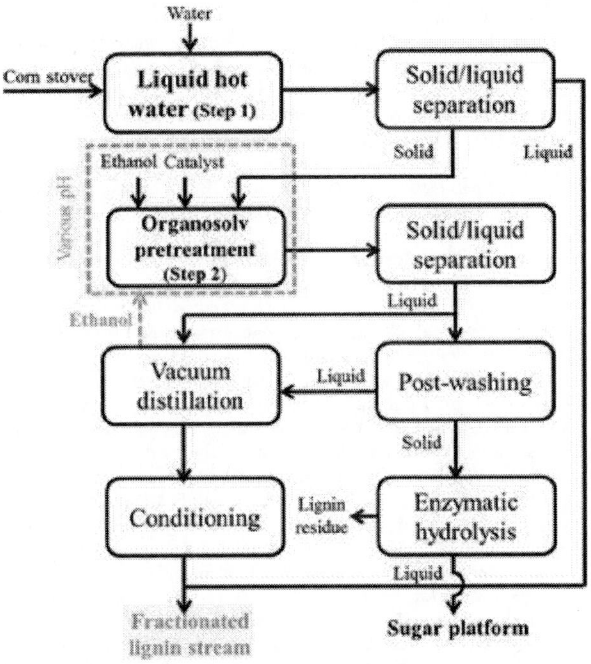

Figure 1. Integrated biomass fractionation process using combinatorial organosolv pretreatment (COP) with controlled pH to produce fermentable sugar and a uniform lignin stream.

■ MATERIAL AND METHODS

Pretreatment Strategies of Corn Stover and Switchgrass. Corn stover and switchgrass used in this study were harvested from Comanche, TX, United States. Five pretreatment strategies were developed to deconstruct corn stover and switchgrass (Table 1). During pretreatment, 50 g of corn stover (dry weight, dw) was loaded into a 1.0-l screw bottle and then pretreated by liquid hot water (step 1) at 120 °C for 30 min in Amsco LG 250 Laboratory Steam Sterilizer (Steris, USA). After that the pretreated slurry was filtered to separate solid from liquid stream. The solid fraction was then treated by using ethanol (EtOH)+formic acid (FA), EtOH+dilute sulfuric acid (SA), or EtOH+sodium hydroxide (SH) (step 2) at 120 °C for 60 min. This fractionation process shown in Figure 1 is named combinatorial organosolv pretreatment (COP).

Pretreatment severity ($\log R_0$) is calculated as follows[32]

$$\log R_0 = t \times \exp[(T - T_b)/\omega] \quad (1)$$

where t is the residence time (min), T is the holding temperature (°C), T_b is the base temperature (100 °C), and ω is the fitted value based on the activation energy (14.75).

The severity factor $\log R_0''$ has been used to compare the severity of COP[18]

$$\log R_0'' = \log R_0 + |\mathrm{pH} - 7| \quad (2)$$

Enzymatic Hydrolysis of Pretreated Solid. The pretreated solids were hydrolyzed by Cellic CTec 2 and HTec 2 commercial enzymes mixtures. Filter paper activity (FPU) of CTec 2 is 96 FPU/mL. The cellobiase activity of CTec 2 is 1270 CBU/mL. The protein content is 178 ± 19.9 mg/mL for Cellic CTec2 and 103 ± 9.6 mg/mL for Cellic HTec 2 based on the Bradford assay. The hydrolysis was carried out in a 250 mL Erlenmeyer flask with a 100 mL total volume and 3.0% solid loading at a constant pH 4.8, 50 °C, and 200 rpm for 168 h. A cellulase loading of 10 FPU/g solid and a volumetric ratio of CTec2:HTec2 10:1 were used.

Fabrication of Lignin Nanoparticles (LNPs) via Antisolvent Precipitation. Lignin was recovered by filtering of the liquid stream fractionated from each COP with a 0.22 μm filter membrane and acidifying to the pH 2.0 by sulfuric acid. The lignin precipitates were collected by centrifugation at 10 000 rpm for 15 min and then freeze dried at −55 °C for 24 h by a lyophilizer (Labconco Corp., USA). The fractionated lignin without postmodification was used for LNPs fabrication via antisolvent precipitation. In detail, lignin was first dissolved to 10 mg/mL in acetone and water (9:1, v/v) and treated for 30 min by sonication to form a pure solution. The mixture was then filtered through a 0.22 μm syringe filter to remove undissolved particles and rapidly loaded into ddH_2O. The suspension was centrifuged at 1200g for 10 min to remove aggregates, and the dispersion was used to determine the size and morphology of LNPs. The LNP yield was calculated based on the lignin used for LNP preparation by using the gravimetric method.

Characterization of Lignin Nanoparticles (LNPs). Transmission electron microscopy (TEM) analysis of LNPs was performed using a ZEISS LIBRA 200 high-tilt tomography transmission electron microscope with an acceleration voltage of 120 kV. The dispersion of LNPs was dropped on a carbon film support grid and incubated for 2 min, and the excess water was removed by blotting the side of the grid onto filter paper. Imaging was performed in a brightfield mode with slight underfocus. The particle size and zeta potential of the LNPs were determined by using a Brookhaven ZetaPlus Zeta Potential Analyzer, Brookhaven Instruments Corp., New York, USA.

Characterization of Fractioned Lignins and Solids. 2D $^1H-^{13}C$ heteronuclear single quantum coherence (HSQC) nuclear magnetic resonance (NMR) analysis of lignin were carried out on a Varian 500 MHz NMR spectrometer with the "gradient HSQCAD" mode. A 30−50 mg amount of lignin samples was dissolved in 0.6 mLof dimethyl sulfoxide-d_6. The gradient-enhanced HSQC with adiabatic pulses mode was used with the following parameters: 1.0 pulse delay, 32 scans, 1024 data points for ^1H, and 256 increments for ^{13}C. For data analysis, the ^1H and ^{13}C spectral widths were 13.0 and 220.0 ppm, respectively. The $\delta C/\delta H$ = 39.5/2.49 ppm of the central solvent peak was used as reference. The identification of ferulate and *p*-coumaric acid is based on the C−H bond signal of the No. 2 carbon and the No. 2/6 carbon, respectively. The semiquantitative calculation is based on the relative percentage of the signal integral.

^{31}P NMR spectra were obtained by previous methods.[33] A 20−25 mg amount of lignin samples was dissolved in 0.7 mL of stock solution of pyridine/$CDCl_3$ (v/v = 1.6/1). The stock solution included 1.25 mg/mL $Cr(acac)_3$ and 2.5 mg/mL internal standard *endo-N*-hydroxy-5-norbene-2,3-dicarboxylic acid imide. After that 70 μL of phosphitylating reagent 2-chloro-4,4,5,5-tetramethyl-1,3,2-dioxaphospholane was added to the vial. Quantitative ^{31}P NMR analysis was conducted on a Varian 500 MHz spectrometer with an inverse-gated decoupling pulse sequence, 90° pulse angle, 1.2 s acquisition time, 25 s pulse delay, and 64 scans.

For gel-permeation chromatography (GPC) analysis, the fractionated lignins were acetylated with acetic anhydride/pyridine (1/1, v/v) at 2 mg/mL of lignin concentration for 24 h. A 20 mL amount of ethanol was then added into the solution, and the mixture was stirred for 30 min. The solvents were evaporated by using a rotary evaporator followed by drying at 40 °C. The acetylated lignin samples were dissolved by tetrahydrofuran at a concentration of 1.0 mg/mL. The molecular weight of lignin was analyzed on an Agilent GPC SECurity 1200 system, which was equipped with an Agilent refractive index (RI) detector, an Agilent UV detector (270 nm), and three Waters Styragel columns (HR1, HR2, and HR6). Tetrahydrofuran was used as the mobile phase, and the flow rate is 1.0 mL/min.

The morphology of pretreated solids was determined by scanning electron microscopy (SEM). Pretreated solids were mounted onto stubs with carbon tape and sputter coated with gold. SEM was carried out on Zeiss EVO MA15 at an accelerating voltage of 20 kV with a back scatter detector at ~50–2000 times magnification. Raw images were adjusted for brightness and contrast in ImageJ software.[34] Images were merged using Adobe Photoshop CC v. 2017.

Analysis Methods. Composition analysis was conducted according to the Laboratory Analysis Protocol (LAP) of the National Renewable Energy Laboratory (NREL), Golden, CO, USA. The sugars were analyzed by HPLC (HPLC 1260 Infinity; Agilent Technologies, CA) equipped with an Aminex HPX-87P carbohydrate analysis column (Bio-Rad Laboratories, CA) and a refractive index detector using HPLC-grade water as the mobile phase at a flow rate of 0.6 mL/min. Sugar yields were calculated based on the initial glucan and xylan content in feedstock.

■ RESULTS AND DISCUSSION

Component Transformation in Combinatorial Organosolv Pretreatment (COP). Component transformation analysis is essential to evaluate the pretreatment performance.[35,36] As shown in Table 2, the total sugar content in corn stover and switchgrass was 51.4% and 56.1%, respectively, while the lignin content was 16.3% and 13.6%. Interestingly, there was approximately 7.3% free glucose in water extractives of corn stover, in contrast to only 2.6% free glucose in switchgrass. Free sugar is more degradable in pretreatment and may generate inhibitors.[37,38] Thus, pretreatment options should consider their effects on sugar and lignin transformation.

Table 2. Compositions of Corn Stover and Switchgrass Biomass Feedstock[a]

components	UCS	USG
glucan	26.8 (1.1)	34.3 (2.2)
xylan	15.0 (0.5)	17.7 (0.8)
arabinan	1.0 (0.2)	0.6 (0.1)
galactan	0.4 (0.0)	0.4 (0.0)
mannan	0.5 (0.0)	0.4 (0.0)
lignin		
-acid-insoluble lignin (ail)	13.5 (1.7)	11.9 (0.1)
-acid-soluble lignin (asl)	2.8 (0.1)	1.7 (0.1)
water extractives	26.4 (5.4)	9.8 (1.2)
-free glucose	7.3 (0.4)	2.6 (0.1)
-free xylose	0.4 (0.1)	0.1 (0.0)
ethanol extractives	5.0 (0.1)	3.4 (1.1)
ash	8.7 (0.6)	3.4 (0.2)
total	99.8	83.5

[a]Note: UCS represents untreated corn stover; USG represents untreated switchgrass; all data in the table are mean values of duplicate experiments; Standard deviation is given in parentheses.

Combinatorial organosolv pretreatment (COP) with a controlled pH value was designed to simultaneously improve the yields of sugar and lignin in an integrated biorefinery (Table 1 and Figure 1). Figure 2 and Table S1 show that the component transformation in pretreated solid is dependent on the pH value and severity of COP employed. Glucan content decreased from 57.5% to the lowest value of 34.1% as the pH increased from 1.8 to 6.2 and then increased to 48.2% at pH 13.0 (Figure 2A). The results suggested that COP at lower or higher pH enriched glucan content in the pretreated solid compared with the feedstock. Interestingly, glucan content changed little below a severity factor of 8.4 but rapidly increased from 40.5% to 57.5% when the severity factor increased to 11.2 (Figure 2B). COP with a higher severity factor enriched glucan content. Xylan content increased from 11.1% to 21.5% with increasing pH (Figure 2C). Arabinan showed the same trend as xylan (Figure 2E). Generally, pretreatment at low pH tends to degrade and remove xylan and arabinan via the autohydrolysis effect.[4,39] Notably, the correlations of xylan and arabinan content with the severity factor were not significant (Figure 2D and 2F), indicating that xylan and arabinan content were more sensitive to the low pH of COP. The lignin content increased from 10.5% to 18.5% as the pH increased from 1.8 to 6.0 and then decreased to the lowest value of 9.4% at pH 13.0 (Figure 2G). The results indicated that COP employed lower or higher pH may dissolve more lignin into the liquid stream. The lignin content changed little below a severity factor of 7.8 but then decreased from 18.5% to 9.4% (Figure 2H). These results suggested that glucan and lignin contents in the pretreated solid were influenced by both pH and severity of COP, whereas xylan and arabinan were more sensitive to pH. Overall, component transformation was heavily dependent on the COP employed and the severity and pH corresponding to it.

Enzymatic Hydrolysis of the Pretreated Solid Produced from Each COP. Enzymatic hydrolysis of the pretreated solid was carried out to evaluate the efficiency of the sugar platform. Figure 3 shows that COPs using EtOH+SA and EtOH+SH resulted in more than 95% glucan and 96% xylan conversion, indicating the improved hydrolysis performance. Figure 4 shows that glucan conversion rapidly increased from 40.1% to 95.6% and xylan conversion increased from 37.6% to 96.5% as the severity factor increased. COPs with lower/higher pH or high severity deconstructed the rigid matrix of corn stover/switchgrass by dissolving more lignin and xylan. These modifications should lead to the improved accessibility of cellulose and hemicellulose to enzymes and thus increased the hydrolysis performance.

The scanning electron microscopy (SEM) images in Figure S1 show the macrostructure change of corn stover/switchgrass depended on the COP employed. The overall retention of shape in the lower magnification images (~50×) and the presence of intact cell walls in the higher-magnification images (500× and 2000×) are likely due to a low holding temperature employed in COP. However, COP using EtOH+SA significantly changed the morphology of corn stover, which was much different from that obtained using EtOH or EtOH+FA (Figure S1 A–D). COP using EtOH+SA led to clumps

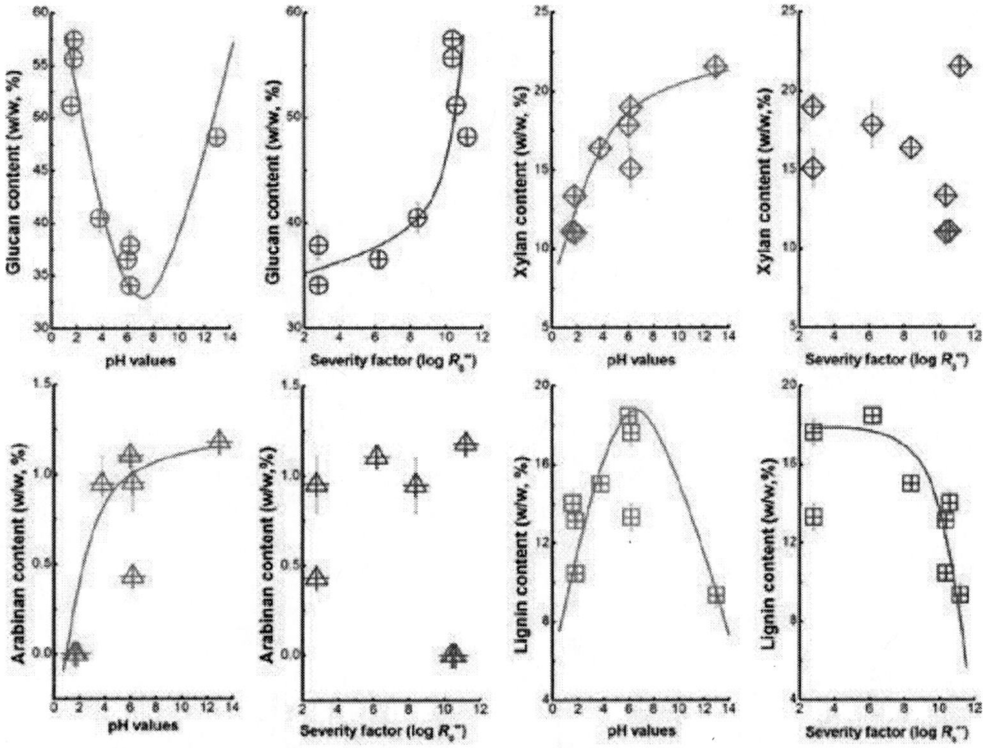

Figure 2. Component transformation in the pretreated solid by combinatorial organosolv pretreatment (COP) at various pH values and severity factors

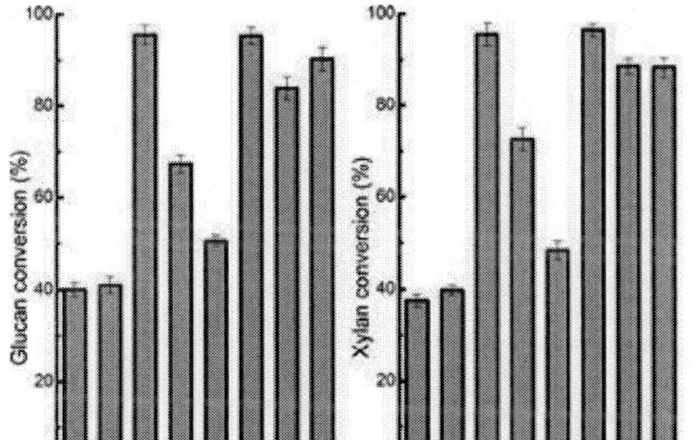

Figure 3. Enzymatic hydrolysis of the pretreated solid produced from combinatorial organosolv pretreatment (COP) with controlled pH. LHW represents liquid hot water; EtOH+SA represents ethanol plus sulfuric acid; EtOH+FA represents ethanol plus formic acid; EtOH +SH represents ethanol plus sodium hydroxide; MeOH+SA represents methanol plus sulfuric acid; Feedstock used in LHW* and EtOH+SA* is switchgrass.

due to the intertwining of bulk fibers. Individual fibers emerging from the bulk fibers were observed, suggesting significant breakdown of the macrostructure, which facilitated the improvement of hydrolysis performance. COP using MeOH+SA caused changes in corn stover similar to those of EtOH+SA (Figure S1F). In contrast, COP using EtOH+SH caused a significant change in the macrostructure of corn stover that differed from those caused by using EtOH+SA or MeOH+SA (Figure S1E). Although a large deformation was caused by using EtOH+SH, few individual fiber strands emerged. COP using EtOH+SA caused intertwining of switchgrass, with subsequent formation of clumps and emergence of individual fibers from the particle bulk, similar to that observed in corn stover (Figure S1G). These effects also facilitated the improvement of hydrolysis performance. Overall, COP using EtOH+SA/SH modified and deconstructed the macrostructure of corn stover/switchgrass, resulting in improved hydrolysis performance.

The sugar yield is a crucial metric for evaluating pretreatment efficacy and will determine biorefinery performance. Figure 5 shows that sugar yield is also dependent on the pH value and severity of COP employed. The glucose yield from corn stover was highest for COP using EtOH+SA (90.1%), followed by EtOH+SH (89.7%), EtOH+FA (74.5%), and EtOH (63.9%). The xylose yield reached the maximum value in COP using EtOH+SH (76.5%), followed by EtOH+SH (73.2%), EtOH+FA (62.8%), and EtOH (50.4%). COP using MeOH+SA produced 86.3% glucose and 66.2% xylose yield from corn stover, while COP using EtOH+SA produced 80.2% glucose and 72.8% xylose yield from switchgrass. Results also showed that higher sugar yield was produced from COP under lower or higher pH values (Figure S2), while sugar yield increased slightly below a severity factor of 6.2 but then rapidly increased as the severity factor increased from 6.2 to 11.2. Therefore, COP using EtOH+SA/SH obviously improved the sugar release from biomass, facilitating biorefinery performance.

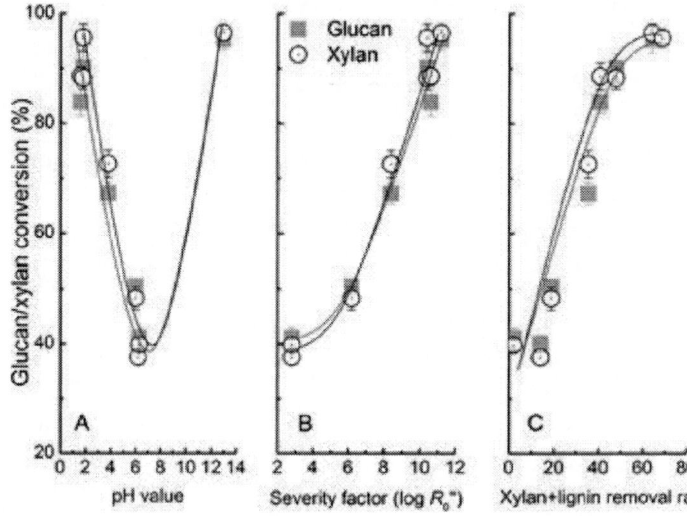

Figure 4. Correlations of glucan/xylan conversion with the pH value (A) and severity factor (log R_0'') (B) of combinatorial organosolv pretreatment (COP) and the removal rates of xylan+lignin (C) in corn stover/switchgrass.

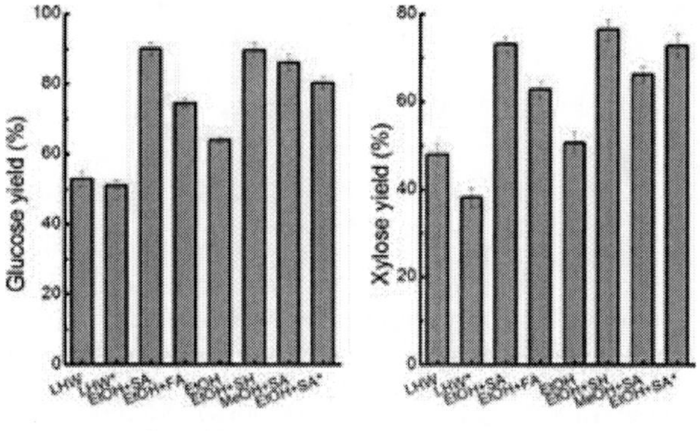

Figure 5. Sugar yield from corn stover/switchgrass subjected to combinatorial organosolv pretreatment (COP). LHW represents liquid hot water; EtOH+SA represents ethanol plus sulfuric acid; EtOH+FA represents ethanol plus formic acid; EtOH+SH represents ethanol plus sodium hydroxide; MeOH+SA represents methanol plus sulfuric acid. Feedstock used in LHW* and EtOH +SA* represents switchgrass.

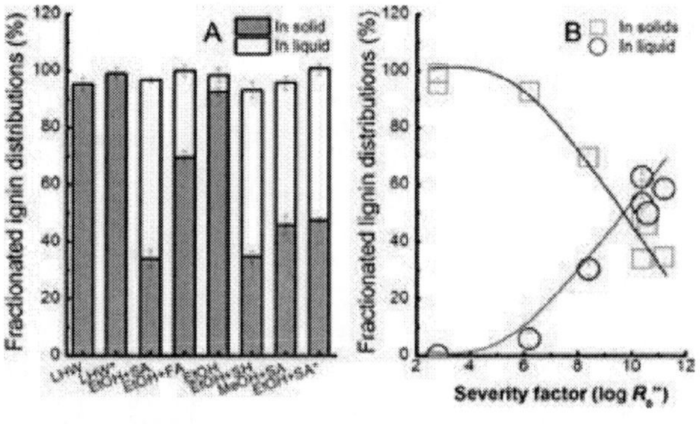

Figure 6. Lignin distributions in solid and liquid streams (A) and their correlations with the severity factor of combinatorial organosolv pretreatment (COP) (B). LHW represents liquid hot water; EtOH +SA represents ethanol plus sulfuric acid; EtOH+FA represents ethanol plus formic acid; EtOH+SH represents ethanol plus sodium hydroxide; MeOH+SA represents methanol plus sulfuric acid. Feedstock used in LHW* and EtOH+SA* is switchgrass.

Improved Lignin Fractionation Efficiency by COP. Lignin valorization to produce value-added products provides additional revenue for a sustainable biorefinery. The lignin yields are crucial for its valorization and thus biorefinery economy.[40−42] Figure 6A shows that the lignin distribution in the solid and liquid streams depended greatly on the COP employed. Approximately 93% and 70% of the lignin was retained in the solid residue produced from COP using EtOH and EtOH+FA, respectively, likely due to the low pretreatment severity employed. However, only 34−46% of the lignin was retained in the solid residue produced from COP using EtOH +SA/SH. Correspondingly, COP using EtOH+SA and EtOH +SH dissolved approximately 62.3% and 58.5% of the lignin into the liquid stream, respectively, suggesting the high fractionation efficiency. Figure 6B shows the correlations between lignin distributions and the severity factor of COP. Most of the lignin was retained in the solid fraction at severity factors of less than 6.2, and the content decreased significantly as the severity factor further increased. Correspondingly, the lignin yield in the liquid stream increased sharply as the severity factor of COP increased from 6.2 to 11.2. Overall, COP using EtOH+SA/SH deconstructed corn stover/switchgrass, depolymerized and dissolved more lignin into the liquid stream, and thus improved lignin output for its further valorization.

Fabrication of Lignin Nanoparticles (LNPs) via Antisolvent Precipitation. The fractionated lignin without postmodification was used to assess the potential of LNP fabrication via antisolvent precipitation. Lignin is an amphiphilic polymer formed from hydrophobic phenylpropanoid units with hydrophilic hydroxyl and carboxyl groups. During the fabrication process of LNPs, the hydrophobic aromatic skeletons of lignin likely aggregate in the water to form the cores of particles, while the hydrophilic groups may form the shells of the particles. The images in Figure 7 confirmed that spherical nanosized precipitates with smooth surface were fabricated. The LNPs were symmetric, with a small particle size and uniform round shape. The results

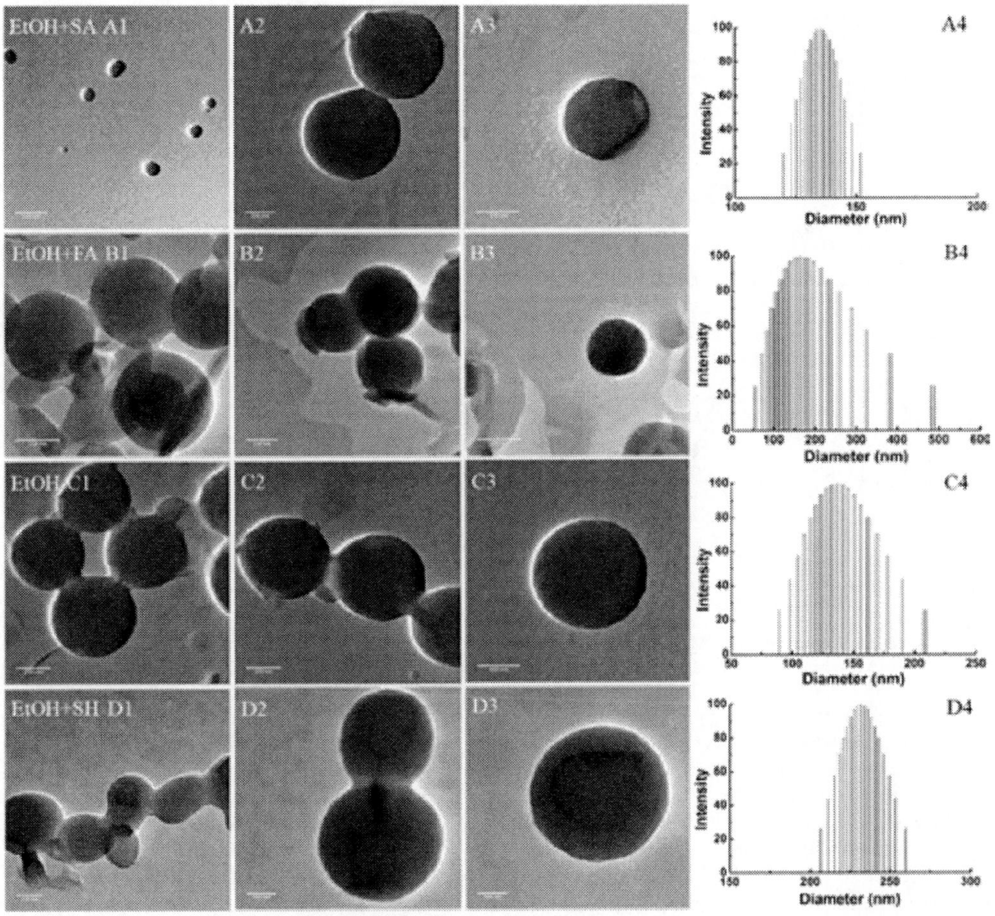

Figure 7. Transmission electron microscopy (TEM) images (A1−3, B1−3, C1−3, D1−3) and particle size distribution (A4−D4) of lignin nanoparticles (LNPs) fabricated via antisolvent precipitation of the lignin fractionated by combinatorial organosolv pretreatment (COP) with controlled pH. EtOH+SA represents ethanol plus sulfuric acid; EtOH+FA represents ethanol plus formic acid; EtOH+SH represents ethanol plus sodium hydroxide.

also showed that some LNPs possessed small aggregate-like structures, likely due to hydrogen bonding between particles. Figure S3 shows that the LNP yield depended on the COP employed. The LNP yield followed the order EtOH+FA (75.8%), EtOH+SH (80.3%), EtOH (82.6%), and EtOH+SA (87%), suggesting that higher LNP yield was obtained by COP using EtOH+SA.

Figure 8 shows that the properties of the LNPs were significantly dependent on the COP employed. The effective diameter of the LNPs determines their properties and applications. The effective diameter followed the order EtOH +SA (142 nm), EtOH+FA (184 nm), EtOH (197 nm), EtOH +SH (234 nm). Results suggested that the lignin fractionated from corn stover by COP can be used as feedstock to produce LNPs with satisfied effective diameter via antisolvent precipitation, while COP with acidic conditions promoted the fabrication of LNPs with a smaller particle size. Previous study reported that the LNPs fabricated from dissolved poplar wood lignin by *p*-toluenesulfonic acid had an average diameter that ranged from 349.7 to 467.1 nm.[43,44] The LNPs obtained by dissolving softwood Kraft lignin in tetrahydrofuran had average particle sizes between 320 and 360 nm.[25] A Gaussian function was employed to fit the particle size distribution, and the half width of the fitting curve was given to evaluate the uniformity of the LNPs. The half width was less than 50 nm for all LNPs except for that from COP using EtOH+FA. Polydispersity index (PDI) is usually used as one of key metrics to assess the uniformity of particles. The PDI of all LNPs except from COP using EtOH+FA was less than 0.1. Previous research reported that the PDI of LNPs obtained from dissolved poplar wood lignin by *p*-toluenesulfonic acid was 0.163−0.184,[43] which is larger than most of the COP-based LNPs. These results indicated that the LNPs prepared from the lignin fractionated by COP especially at lower pH values exhibited smaller effective diameters and more uniform distributions.

The zeta potential is crucial metric used to evaluate the stability of particle dispersion. Figure 8 shows that the LNPs from all COPs had negative values of zeta potential, likely due to the true negative charges of the phenol groups and the adsorption of hydroxyl ions on the hydrophobic surface of lignin.[25,45] The surface charge could electrostatically stabilize the LNPs to prevent their aggregation in aqueous solution. The zeta potential of LNPs from COP followed the order of EtOH (−18.1 mV), EtOH+FA (−19.9 mV), EtOH+SH (−47.1 mV), and EtOH+SA (−50.8 mV). LNPs with a high

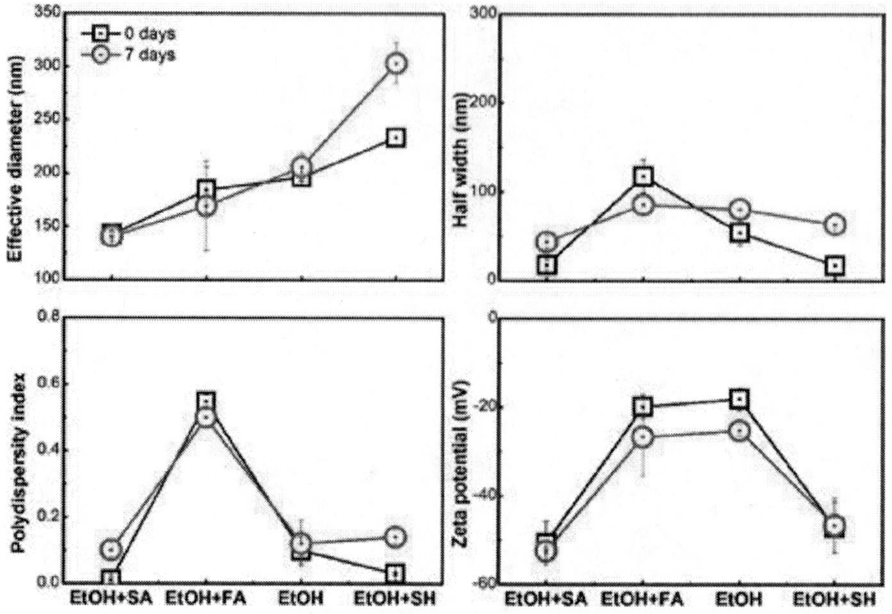

Figure 8. Properties of lignin nanoparticles (LNPs) from the lignin produced by each combinatorial organosolv pretreatment (COP) with controlled pH. EtOH+SA represents ethanol plus sulfuric acid; EtOH+FA represents ethanol plus formic acid; EtOH+SH represents ethanol plus sodium hydroxide.

absolute value of zeta potential are electronically stabilized, while those with a low absolute value of zeta potential are likely to aggregate and have poor stability. Previous studies reported that the zeta potential of LNPs from dissolved poplar wood lignin by *p*-toluenesulfonic acid and low-sulfonated lignin was between −27.2 and −40.0 mV.[43,44,46] LNPs produced from lignin of steam-pretreated corn stover, hardwood poplar, and softwood lodgepole pine gave zeta potentials of approximately −16, −37, and −45 mV, respectively.[45] These results indicated that the COP configurations especially at lower or higher pH can produce a suitable lignin stream to facilitate the fabrication of LNPs, which presented higher zeta potentials and should possess excellent physical stability. Overall, taking these metrics into consideration, COP using EtOH+SA produced more uniform and stable LNPs with a relatively smaller effective diameter, lower PDI, and higher zeta potential, which may suggest superior performance for their applications.

Enhanced Stability of Lignin Nanoparticles (LNPs) by COP. It is crucial that LNPs retain their properties under certain conditions for further application, such as in drug delivery, functional surface coatings, and slow-release fertilizers, etc. Thus, the stability of the LNPs was evaluated as a function of storage period, pH, and salt concentration. Figure 8 illustrates the effects of storage period on the stability of the LNPs. The effective diameter, half width, PDI, and zeta potential of the LNPs barely changed after 7 days, and no specific aggregation of synthesized LNPs was observed in aqueous dispersion. Results suggested the long-term stability of the LNPs and the potential for various applications. This stability may be attributed to the electrostatic repulsion induced by the high surface charge of these LNPs.[26]

The pH value of dispersion is a key variable affecting the properties and applications of LNPs, and thus, the stability of the LNPs was evaluated at various pH values (Figure S4). The results showed that the properties of the LNPs depended on the pH value of the aqueous dispersion. The effective diameter at pH 3.0 was 1.3−1.8 times higher than that under neutral conditions and slightly increased at pH 11.0. These results suggested that LNPs at pH 3.0 tended to aggregate. The half width and PDI significantly increased at pH 3.0 and 11.0, suggesting reduced uniformity of the LNPs. The electrical double-layer repulsion resulting from the hydroxyl and carboxyl groups was responsible for the stability of LNPs.[45] The zeta potential decreased by 30−35% at pH 3.0 compared with neutral conditions, indicating weak electrostatic repulsion between the particles. However, at pH 11.0, the zeta potential changed little for COP using EtOH+SA and EtOH+SH and increased by 62% and 67% for COP using EtOH+FA and EtOH, respectively, suggesting strong stability of these LNPs. Previous studies reported that LNPs prepared from steam-exploded lignin showed good stability between pH 4.0 and 10.0 but started to aggregate beyond these pH values.[45] In addition, LNPs prepared from organosolv lignin showed good stability in the pH range of 3.5−8, but aggregation of LNPs was observed below pH 3.2.[26] The protonation and deprotonation reactions of hydrophilic hydroxyl and carboxyl groups may be the main reason for these results.[25,45] At low pH, the LNPs were likely to aggregate because of the reduced electrical layer repulsion induced by the protonation of charged functional groups. However, the LNPs tended to disaggregate at high pH and showed a higher effective diameter. Taken together, although the PDI increased at pH 3.0 and 11.0, considering the high zeta potential, the LNPs fabricated from the COP fractionated lignin still showed high stability at a broad range of pH values.

The sensitivity of the LNPs to changes in salt concentration was evaluated (Figure S5). Although a slight reduction in the zeta potential was observed at 100 mM NaCl, the LNPs was still stable with no significant increase in the effective diameter, half width, or PDI. When the concentration of NaCl increased to 500 mM, the effective diameter, half width, and PDI of the LNPs significantly decreased, and the zeta potential almost

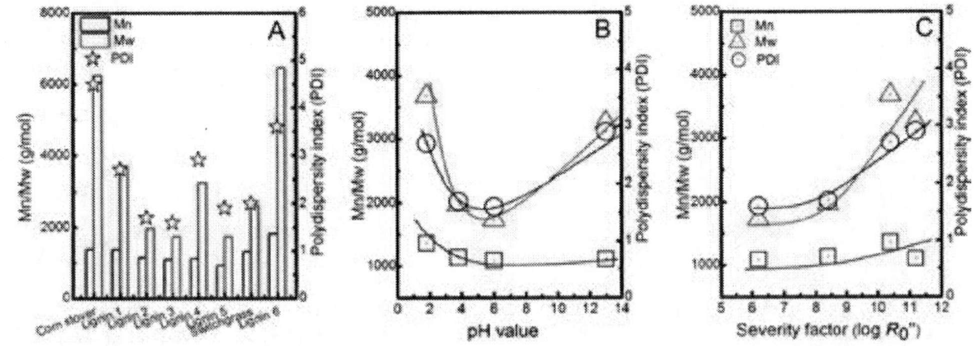

Figure 9. Molecular weight distribution of the lignin fractionated by combinatorial organosolv pretreatment (COP) with controlled pH. M_n represents the number-average molecular weight; M_w represents the weight-average molecular weight. Lignin 1 represents the fractionated lignin by COP case 1 described in Table 1.

reached 0 mV. These results suggested that the LNPs started to aggregate at 500 mM NaCl. The aggregation of LNPs and the reduction of zeta potential were also observed with increasing salt concentration in previous studies.[25,26] The increase in salt concentration may reduce the range of double-layer repulsion between particles, eventually resulting in their aggregation. The accumulation of Na^+ around the LNPs also led to a reduction of the thickness of the electrical double layer and thus a decrease in the zeta potential. Taken together, these results suggested that the LNPs from COP possessed stability and uniformity within a specific salt concentration.

Tuned Lignin Chemistry by COP for LNP Fabrication. The lignin chemistry determines the reactivity of lignin and thus the properties of LNPs. The pretreatment can tune the lignin chemistry, regulate the lignin processability, and thus determine the properties of LNPs. To reveal the mechanism by which COP modifies lignin chemistry and its correlation with the LNP properties, 2D and ^{31}P NMR and GPC were employed to characterize the chemistry of lignin.

Molecular Weight of the Fractionated Lignin from COP. Figure 9 shows that the molecular weight of fractionated lignin obviously depended on the COP employed. The native lignin of corn stover showed a number-average molecular weight (M_n) of 1371 g/mol and a weight-average molecular weight (M_w) of 6241 g/mol (Figure 9A). For lignins 1−5 from corn stover, M_n ranged from 916 to 1363 g/mol and M_w ranged from 1732 to 3679 g/mol, suggesting depolymerization of lignin by COP. The detailed results showed that the M_n of the fractionated lignin slightly decreased with increasing COP pH, which was opposite the trend of the changes in the effective diameter of the LNPs. However, the M_w of the fractionated lignin rapidly decreased as the COP pH increased from 1.8 to 6.2 and then increased with increasing pH (Figure 9B). Lignin 1 produced from corn stover by EtOH+SA showed higher M_n and M_w compared with other COPs, possibly due to acid-catalyzed condensation of lignin. A similar effect was observed for switchgrass, as lignin 6 showed a considerably higher molecular weight than the native lignin of switchgrass. Previous studies have confirmed that under acidic conditions the fractionation process of lignin is the result of competition between depolymerization and condensation.[47−49] Generally, a low severity of pretreatment produced a greater extent of interlinkage cleavage of lignin, leading to a decrease in M_w, while with increasing severity, the condensation reaction became more dominant, resulting in an increase in lignin molecular weight. This condensed lignin produced from acid fractionation should enhance the hydrophobicity of lignin and facilitate the fabrication of LNPs with small particle size (Figure 8).[45]

Compared with native lignin, COP fractionated lignin with lower PDI and improved uniformity of lignin stream, which facilitate the upgrading of lignin to LNPs (Figure 9). The results also showed higher PDI values for lignins 1 (PDI = 2.7), 4 (PDI = 2.9), and 6 (PDI = 3.6) compared with other lignins, which indicated a wider molecular weight distribution and obvious depolymerization of lignin by COP at lower or higher and suggested the improved processability. As shown in Figure 9C, M_w, M_n, and PDI changed little below a severity factor of 5.6 and then increased with increasing severity factor. COP using EtOH+FA and EtOH produced lignin with lower M_n and M_w, probably due to extraction of the easily dissolved lignin with a smaller molecular weight,[17] as supported by lignin distribution analysis (Figure 6).

Changes in Lignin Subunits by COP. Figure 10 presents the results for each compositional unit from the 2D HSQC NMR spectra (Figure S6 and Figure 10A). The fractionated lignin subunits were significantly dependent on the COP employed. The S-lignin content increased with increasing pH, showing a similar trend to the effective diameter of the LNPs (Figure 10B). G-lignin significantly decreased with increasing pH, which showed an opposite trend to the effective diameter. COP using EtOH+SA decreased S-lignin and increased G-lignin content compared with native lignin, which may facilitate the LNP fabrication with a smaller effective diameter, while COP using EtOH+SH increased S-lignin and decreased G-lignin content, accounting for the increased effective diameter of LNPs. The correlation between lignin unit type and severity factor of COP is fit in Figure 10C but was not significant. The results suggested that the lignin subunits were more sensitive to pH than to the severity factor of COP. As shown in Figure 10B, the S/G ratio in fractionated lignin significantly increased with increasing pH, similar to the trend for the effective diameter of the LNPs. The change in S/G ratio with COP severity factor is fit in Figure 10C but was not significant. A positive relation between the S-unit content of lignin from steam-exploded corn stover and hardwood poplar and the particle size of LNPs was reported.[45] Therefore, the higher G-unit and lower S-unit content in lignin should facilitate the fabrication of LNPs with a small particle size, as supported by the results for the LNPs (Figure 8).

The interunit linkage results showed that COP using EtOH+SA produced a lower content of β-O-4 linkages in lignin

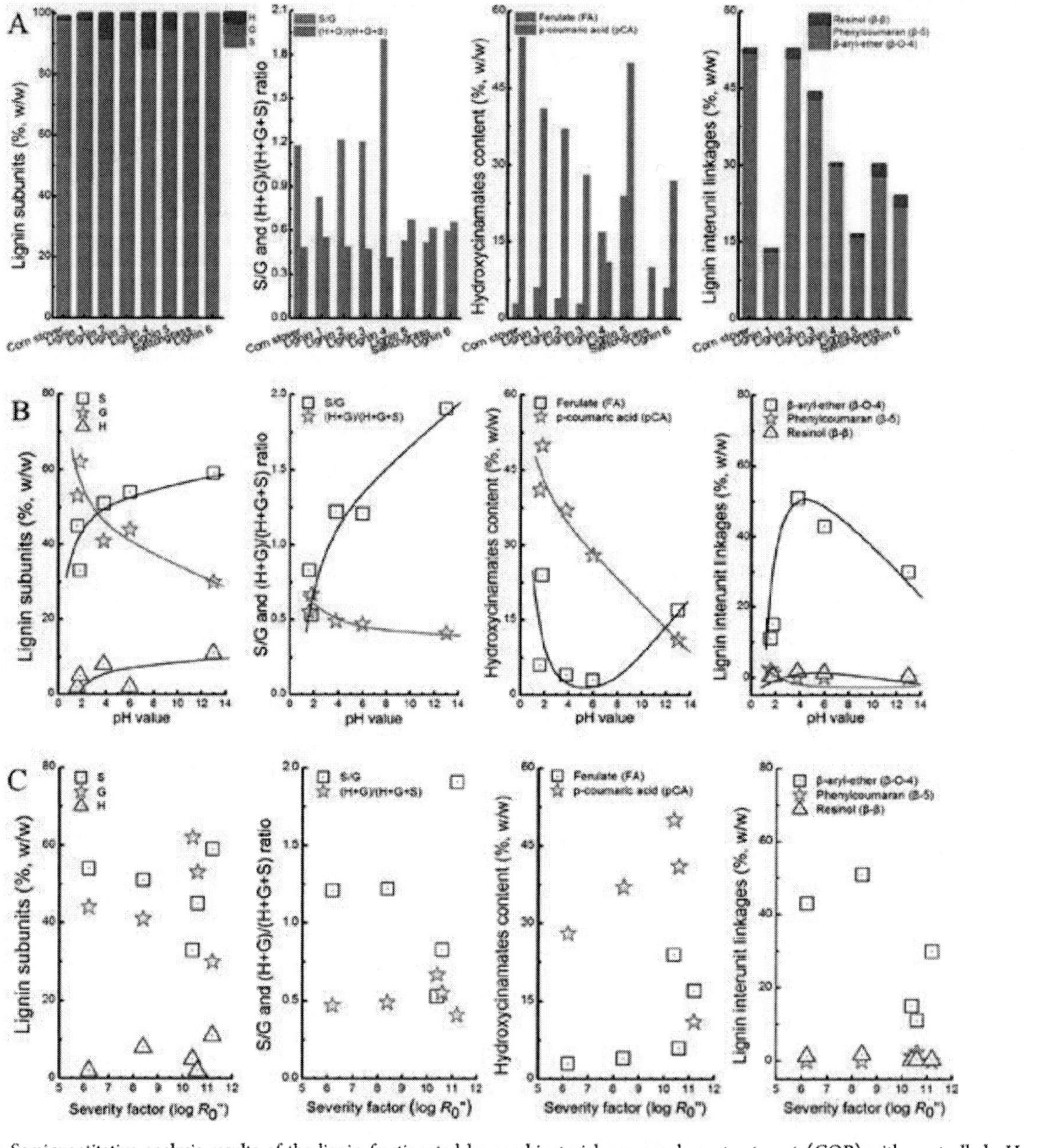

Figure 10. Semiquantitative analysis results of the lignin fractionated by combinatorial organosolv pretreatment (COP) with controlled pH as detected using 2D HSQC NMR. Lignin 1 represents the fractionated lignin by COP case 1 described in Table 1.

(lignin 1) compared with other COPs (Figure 10A). The cleavage of *β*-O-4 linkages by COP using EtOH+SA did not result in a lower molecular weight of lignin (Figure 9), supporting the formation of condensed lignin. The contents of *β*-O-4 linkages increased as the pH increased from 1.8 to 6.0 and then decreased with further increases in pH. The results indicated that COPs with both acidic and alkaline conditions were effective to cleave the *β*-O-4 linkages, increase the depolymerization of lignin, and thus expose more functional groups (Figure 10B). Notably, the content of *β*-5 linkages was 2.3% in lignin produced by COP using EtOH+SA; however, *β*-5 linkages were not detected in lignin produced from corn stover by other COPs. These results suggested that COP with high pH broke down more *β*-5 linkages. The content of *β*–*β* linkages increased from 0.3% to 1.8% and then decreased to 0.3% with increasing pH, suggesting the cleavage of *β*-5 linkages by COP at lower or higher pH. Overall, more cleavage of interunit linkage by COP especially using EtOH+SA indicated obvious depolymerization of lignin and should expose more functional groups to improve lignin processability for LNP fabrication.

Hydroxyl Groups in Lignin Fractionated by COP. As the hydrophilic groups promote the formation of electrical double layers and determine the formation and properties of LNPs, the tuned hydroxyl groups in fractionated lignin by each COP were determined by ^{31}P NMR.[33,50] All COPs increased

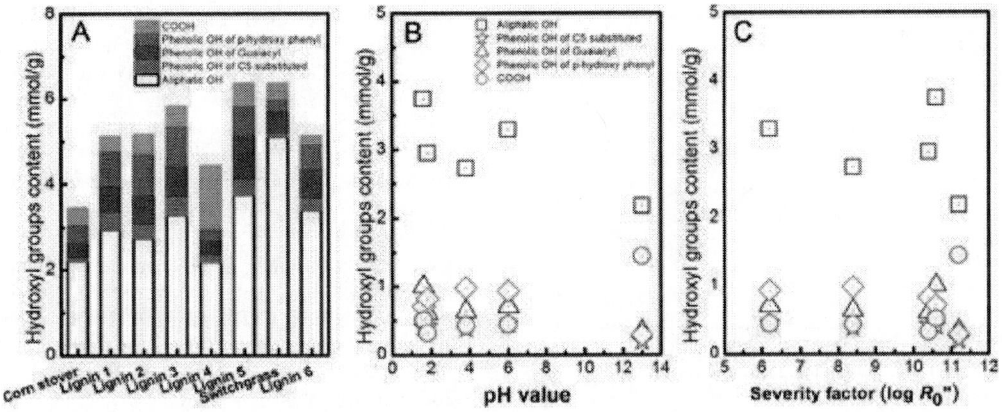

Figure 11. Hydroxyl group content in the lignin fractionated by combinatorial organosolv pretreatment (COP) with controlled pH as detected using ^{31}P NMR. Lignin 1 represents the fractionated lignin by COP case 1 described in Table 1.

the aliphatic OH content in fractionated lignin compared with native lignin and led to the depolymerization of lignin, supporting the results of interunit linkage cleavage (Figure 11A). The aliphatic OH content decreased slightly with increasing pH, suggesting that COP with acidic conditions was conducive to increasing this group. Interestingly, lignin with higher aliphatic OH facilitated the fabrication of LNPs with a small diameter likely due to the enhanced electrical double layers.

For phenol OH group, the C5-substituted OH content in fractionated lignin was approximately 0.40−0.45 mmol/g below pH 6.2 but decreased to 0.19 mmol/g at pH 11.2. The COPs under acidic conditions enriched the C5-substituted OH, possibly due to the lignin condensation reaction.[51,52] As aforementioned, condensed lignin will improve lignin hydrophobicity to facilitate the LNP fabrication with a small diameter. The lignins produced from COPs using EtOH+SA and EtOH+FA showed higher contents of guaiacyl and *p*-hydroxy phenyl OH units than native lignin; however, lignin 4 produced from COP using EtOH+SH showed an opposite trend (Figure 11A). Interestingly, the guaiacyl and *p*-hydroxy phenyl OH content in lignin fractionated by COP increased as the pH increased from 1.8 to 6.2 and then decreased with further increases in pH (Figure 11B). Both guaiacyl and *p*-hydroxy phenyl OH content decreased with increasing COP severity factor (Figure 11C). However, compared with alkaline conditions, COP with acidic conditions generated more total phenolic OH groups, especially C5-substituted OH. This result may enhance the stability of LNPs likely due to the formation of intramolecular hydrogen-bonding networks within LNPs because the planar phenol groups in fractionated lignin may be densely packed to form layered structure within the core of LNPs.

COOH groups possibly confer a particle surface charge that promotes the formation of electrical double layers and stabilizes the LNPs. The COOH group content after COP was 1.2−5.5 times higher than that in native lignin, which was consistent with the results of interunit linkage cleavage (Figure 11A). The COOH group content in fractionated lignin decreased from 0.51 to 0.44 mmol/g as the pH value increased from 1.8 to 6.2 and then increased to 1.44 mmol/g at pH 11.2 (Figure 11B). By comparison, the COOH group content increased with the COP severity factor. These results suggested that COP with lower/higher pH or higher severity of pretreatment remarkably enriched the COOH group content in fractionated lignin. Therefore, by increasing the COOH group as well as total phenol OH groups, both COPs using EtOH+SA and EtOH+SH should promote the formation of electrical double layers of LNPs, confirming by the zeta potential of the LNPs (Figure 8), and hence stabilizes the LNPs.

Improved Biorefinery Sustainability by Coproducing Sugar and LNPs. In an integrated biorefinery, COP with controlled pH was designed to produce the sugar platform and fractionate uniform lignin to facilitate high-quality LNP fabrication (Figure 1). COPs developed in this study could offer several significant advantages. COP with lower or higher pH improved the release of sugar from biomass and thus increased the sugar yield. COP using EtOH+SA and EtOH+SH produced 90.1% and 89.7% glucose yield and 73.2% and 76.5% xylose yield, respectively. The liquid hot water employed in step 1 of the COP removed free sugar, water extractives, ash, and other impurities, thereby reducing the formation of potential degradation product and purifying the sugar and lignin streams. Next, ethanol using different catalysts was then employed in step 2 to further deconstruct the corn stover/switchgrass, fractionate the lignin, and thus improve the hydrolysis performance. In addition, COP employing a low holding temperature (120 °C) should reduce sugar degradation and the need for energy consumption compared with conventional hydrothermal pretreatments carried out at temperatures greater than 180 °C.[4,8] Furthermore, the ethanol solvent can be easily recycled and reused.[17]

COP design also improved lignin processability and facilitated the lignin valorization by fabricating high-quality LNPs, which contributed to the sustainability of the biorefinery (Figures 12 and Figure S7). First, despite increasing the cleavage of *β*-O-4 and *β*−*β* linkages, COP with acidic conditions such as using EtOH+SA produced lignin with a higher molecular weight compared with alkaline conditions, likely due to the lignin condensation reaction (Figure 12). COP with acidic conditions also enriched C5-substituted OH content, further confirming the condensation reaction. Because the covalent carbon−carbon single bonds are stronger than van der Waals interactions, lignin condensation among aromatics potentially enhanced the hydrophobic interaction of fractionated lignin during the fabrication of LNPs and thus generated the LNPs with a small particle size.[30,45] Second, COP under

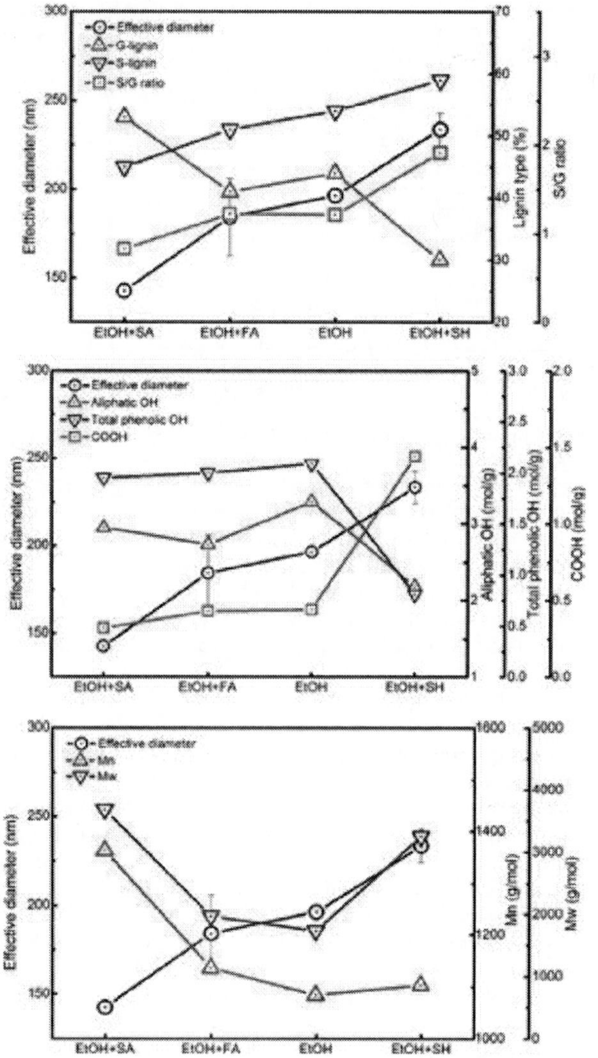

Figure 12. Correlations of the properties of lignin nanoparticles (LNPs) with tuned lignin chemistry by combinatorial organosolv pretreatment (COP). EtOH+SA represents ethanol plus sulfuric acid; EtOH+FA represents ethanol plus formic acid; EtOH+SH represents ethanol plus sodium hydroxide.

acidic conditions produced less S-lignin and higher G-lignin content and thus decreased the S/G ratio, which may facilitate the formation of LNPs with smaller diameters. Third, COP with acidic conditions enriched the total phenolic OH content. Because planar phenol groups of the fractionated lignin may be packed densely as layered structure within the LNPs, the increased total phenolic OH groups should enhance the stability of LNPs with a small effective diameter through the formation of intramolecular hydrogen-bonding networks. By contrast, the LNPs produced from COP with alkaline conditions presented a higher effective diameter, likely due to the enriched S- and H-lignin content, the increased S/G ratio, and the lower total phenolic OH content in the fractionated lignin. Fourth, COP with lower or higher pH produced a higher content of COOH groups in the fractionated lignin. Together with increased total phenolic OH content, they promoted the formation of electrical double layers and increased the zeta potential of LNPs, which improved the stability of LNPs by electrostatic repulsion. Overall, the COP, especially using EtOH+SA, improved the susceptibility and processability of lignin to fabricate high-quality LNPs with more uniform and stable properties due to a smaller effective diameter, a lower PDI, and a higher zeta potential, which should help improve biorefinery sustainability.

■ CONCLUSIONS

COP with controlled pH has been designed to establish a sugar platform and tune lignin chemistry for LNPs. The glucose and xylose yields from COP using EtOH+SA were more than 90% and 73%, respectively. COP with acidic conditions cleaved more β-O-4 and β–β linkages, enriched the G-lignin content, and increased the total phenolic OH groups, which facilitated the LNP fabrication with more uniformity and stability. Therefore, codesign of COP and LNPs improved the sugar release and enhanced the lignin processability for high-quality LNPs with satisfactory property, which showed a potential avenue to make a sustainable biorefinery.

■ ASSOCIATED CONTENT

Supporting Information

The Supporting Information is available free of charge on the ACS Publications website at DOI: 10.1021/acssuschemeng.8b05715.

Compositions in pretreated solid from COP; SEM images of untreated and pretreated corn stover and switchgrass; correlations of glucose/xylose yield with pH value and severity factor of COP; yield of LNPs fabricated from the lignin fractionated by COP; effects of pH value of the dispersion on the stability of LNPs; effects of NaCl concentration on the stability of LNPs; aromatic and lignin interunit regions of 2D HSQC NMR spectra of the fractionated lignin from each COP; proposed formation mechanism model of LNPs from corn stover lignin produced by COP (PDF)

■ AUTHOR INFORMATION

Corresponding Author

*E-mail: syuan@tamu.edu.

ORCID

Somnath Shinde: 0000-0002-0758-3937

Arthur J. Ragauskas: 0000-0002-3536-554X

Joshua S. Yuan: 0000-0002-3837-4214

Author Contributions

Z.H.L. and J.S.Y. designed the study. Z.H.L. carried out the experiments, performed the statistical analysis, and drafted the manuscript. M.O. and K.K. carried out sugar analysis. N.J.H., S.S., S.B., and A.R. characterized the lignin structure. S.B. and J.D. conducted SEM analysis of pretreated solid. Z.H.L. and X.F.K. conducted the characterization of lignin nanoparticles. All authors provided critical inputs to the manuscript and read and approved the final draft.

Notes

The authors declare no competing financial interest.

■ ACKNOWLEDGMENTS

This work was financially supported by the U.S. DOE (Department of Energy,) EERE (Energy Efficiency and Renewable Energy), and BETO (Bioenergy Technology

Office) (grant nos. DE-EE0006112, DE-EE0007104, DE-EE0008250).

REFERENCES

(1) Ragauskas, A. J.; Beckham, G. T.; Biddy, M. J.; Chandra, R.; Chen, F.; Davis, M. F.; Davison, B. H.; Dixon, R. A.; Gilna, P.; Keller, M. Lignin valorization: improving lignin processing in the biorefinery. *Science* **2014**, *344* (6185), 1246843.

(2) Morais, A. R. C.; da Costa Lopes, A. M.; Bogel-Lukasik, R. Carbon dioxide in biomass processing: contributions to the green biorefinery concept. *Chem. Rev.* **2015**, *115* (1), 3–27.

(3) Himmel, M. E.; Ding, S.-Y.; Johnson, D. K.; Adney, W. S.; Nimlos, M. R.; Brady, J. W.; Foust, T. D. Biomass recalcitrance: engineering plants and enzymes for biofuels production. *Science* **2007**, *315* (5813), 804–807.

(4) Chen, H. Z.; Liu, Z. H. Steam explosion and its combinatorial pretreatment refining technology of plant biomass to bio-based products. *Biotechnol. J.* **2015**, *10* (6), 866–885.

(5) Argyropoulos, D. S.; Crestini, C. A perspective on lignin refining, functionalization, and utilization. *ACS Sustainable Chem. Eng.* **2016**, *4* (10), 5089–5089.

(6) Thakur, V. K.; Thakur, M. K.; Raghavan, P.; Kessler, M. R. Progress in green polymer composites from lignin for multifunctional applications: A Review. *ACS Sustainable Chem. Eng.* **2014**, *2* (5), 1072–1092.

(7) Langan, P.; Petridis, L.; O'Neill, H. M.; Pingali, S. V.; Foston, M.; Nishiyama, Y.; Schulz, R.; Lindner, B.; Hanson, B. L.; Harton, S.; Heller, W. T.; Urban, V.; Evans, B. R.; Gnanakaran, S.; Ragauskas, A. J.; Smith, J. C.; Davison, B. H. Common processes drive the thermochemical pretreatment of lignocellulosic biomass. *Green Chem.* **2014**, *16* (1), 63–68.

(8) Mosier, N.; Wyman, C.; Dale, B.; Elander, R.; Lee, Y. Y.; Holtzapple, M.; Ladisch, M. Features of promising technologies for pretreatment of lignocellulosic biomass. *Bioresour. Technol.* **2005**, *96* (6), 673–686.

(9) Qu, T. J.; Zhang, X. M.; Gu, X. W.; Han, L. J.; Ji, G. Y.; Chen, X. L.; Xiao, W. H. Ball milling for biomass fractionation and pretreatment with aqueous hydroxide solutions. *ACS Sustainable Chem. Eng.* **2017**, *5* (9), 7733–7742.

(10) He, Y. Q.; Zhang, J.; Bao, J. Acceleration of biodetoxification on dilute acid pretreated lignocellulose feedstock by aeration and the consequent ethanol fermentation evaluation. *Biotechnol. Biofuels* **2016**, *9*, 19.

(11) Liu, Z. H.; Chen, H. Z. Biomass-water interaction and its correlations with enzymatic hydrolysis of steam-exploded corn stover. *ACS Sustainable Chem. Eng.* **2016**, *4* (3), 1274–1285.

(12) Sannigrahi, P.; Kim, D. H.; Jung, S.; Ragauskas, A. Pseudo-lignin and pretreatment chemistry. *Energy Environ. Sci.* **2011**, *4* (4), 1306–1310.

(13) Liu, Z. H.; Qin, L.; Li, B. Z.; Yuan, Y. J. Physical and chemical characterizations of corn stover from leading pretreatment methods and effects on enzymatic hydrolysis. *ACS Sustainable Chem. Eng.* **2015**, *3* (1), 140–146.

(14) Hendriks, A. T. W. M.; Zeeman, G. Pretreatments to enhance the digestibility of lignocellulosic biomass. *Bioresour. Technol.* **2009**, *100* (1), 10–18.

(15) Kim, J. S.; Lee, Y. Y.; Kim, T. H. A review on alkaline pretreatment technology for bioconversion of lignocellulosic biomass. *Bioresour. Technol.* **2016**, *199*, 42–48.

(16) Zhang, Z. Y.; Harrison, M. D.; Rackemann, D. W.; Doherty, W. O. S.; O'Hara, I. M. Organosolv pretreatment of plant biomass for enhanced enzymatic saccharification. *Green Chem.* **2016**, *18* (2), 360–381.

(17) Zhao, X. B.; Li, S. M.; Wu, R. C.; Liu, D. H. Organosolv fractionating pre-treatment of lignocellulosic biomass for efficient enzymatic saccharification: chemistry, kinetics, and substrate structures. *Biofuels, Bioprod. Biorefin.* **2017**, *11* (3), 567–590.

(18) Pedersen, M.; Meyer, A. S. Lignocellulose pretreatment severity - relating pH to biomatrix opening. *New Biotechnol.* **2010**, *27* (6), 739–750.

(19) Liu, Z. H.; Qin, L.; Pang, F.; Jin, M. J.; Li, B. Z.; Kang, Y.; Dale, B. E.; Yuan, Y. J. Effects of biomass particle size on steam explosion pretreatment performance for improving the enzyme digestibility of corn stover. *Ind. Crops Prod.* **2013**, *44*, 176–184.

(20) Galkin, M. V.; Samec, J. S. M. Lignin valorization through catalytic lignocellulose fractionation: a fundamental platform for the future biorefinery. *ChemSusChem* **2016**, *9* (13), 1544–1558.

(21) Das, L.; Li, M.; Stevens, J.; Li, W. Q.; Pu, Y. Q.; Ragauskas, A. J.; Shi, J. Characterization and catalytic transfer hydrogenolysis of deep eutectic solvent extracted sorghum lignin to phenolic compounds. *ACS Sustainable Chem. Eng.* **2018**, *6* (8), 10408–10420.

(22) Sipponen, M. H.; Lange, H.; Ago, M.; Crestini, C. Understanding lignin aggregation processes. a case study: budesonide entrapment and stimuli controlled release from lignin nanoparticles. *ACS Sustainable Chem. Eng.* **2018**, *6* (7), 9342–9351.

(23) Duval, A.; Lawoko, M. A review on lignin-based polymeric, micro- and nano-structured materials. *React. Funct. Polym.* **2014**, *85*, 78–96.

(24) He, X. Y.; Luzi, F.; Yang, W. J.; Xiao, Z. F.; Torre, L.; Xie, Y. J.; Puglia, D. Citric acid as green modifier for tuned hydrophilicity of surface modified cellulose and lignin nanoparticles. *ACS Sustainable Chem. Eng.* **2018**, *6* (8), 9966–9978.

(25) Lievonen, M.; Valle-Delgado, J. J.; Mattinen, M. L.; Hult, E. L.; Lintinen, K.; Kostiainen, M. A.; Paananen, A.; Szilvay, G. R.; Setala, H.; Osterberg, M. A simple process for lignin nanoparticle preparation. *Green Chem.* **2016**, *18* (5), 1416–1422.

(26) Richter, A. P.; Bharti, B.; Armstrong, H. B.; Brown, J. S.; Plemmons, D.; Paunov, V. N.; Stoyanov, S. D.; Velev, O. D. Synthesis and characterization of biodegradable lignin nanoparticles with tunable surface properties. *Langmuir* **2016**, *32* (25), 6468–6477.

(27) Rinaldi, R.; Jastrzebski, R.; Clough, M. T.; Ralph, J.; Kennema, M.; Bruijnincx, P. C. A.; Weckhuysen, B. M. Paving the way for lignin valorisation: recent advances in bioengineering, biorefining and catalysis. *Angew. Chem., Int. Ed.* **2016**, *55* (29), 8164–8215.

(28) Fache, M.; Boutevin, B.; Caillol, S. Vanillin production from lignin and its use as a renewable chemical. *ACS Sustainable Chem. Eng.* **2016**, *4* (1), 35–46.

(29) Savy, D.; Mazzei, P.; Drosos, M.; Cozzolino, V.; Lama, L.; Piccolo, A. Molecular characterization of extracts from biorefinery wastes and evaluation of their plant biostimulation. *ACS Sustainable Chem. Eng.* **2017**, *5* (10), 9023–9031.

(30) Qian, Y.; Deng, Y. H.; Qiu, X. Q.; Li, H.; Yang, D. J. Formation of uniform colloidal spheres from lignin, a renewable resource recovered from pulping spent liquor. *Green Chem.* **2014**, *16* (4), 2156–2163.

(31) Tian, D.; Hu, J. G.; Bao, J.; Chandra, R. P.; Saddler, J. N.; Lu, C. H. Lignin valorization: lignin nanoparticles as high-value bio-additive for multifunctional nanocomposites. *Biotechnol. Biofuels* **2017**, *10*, 192.

(32) Overend, R. P.; Chornet, E.; Gascoigne, J. A. Fractionation of Lignocellulosics by Steam-Aqueous Pretreatments. *Philos. Trans. R. Soc., A* **1987**, *321* (1561), 523–536.

(33) Pu, Y. Q.; Cao, S. L.; Ragauskas, A. J. Application of quantitative P-31 NMR in biomass lignin and biofuel precursors characterization. *Energy Environ. Sci.* **2011**, *4* (9), 3154–3166.

(34) Schneider, C. A.; Rasband, W. S.; Eliceiri, K. W. NIH Image to ImageJ: 25 years of image analysis. *Nat. Methods* **2012**, *9* (7), 671–675.

(35) Corma, A.; Iborra, S.; Velty, A. Chemical routes for the transformation of biomass into chemicals. *Chem. Rev.* **2007**, *107* (6), 2411–2502.

(36) Liu, Z. H.; Qin, L.; Zhu, J. Q.; Li, B. Z.; Yuan, Y. J. Simultaneous saccharification and fermentation of steam-exploded corn stover at high glucan loading and high temperature. *Biotechnol. Biofuels* **2014**, *7*, 167.

(37) Liu, Z. H.; Qin, L.; Jin, M. J.; Pang, F.; Li, B. Z.; Kang, Y.; Dale, B. E.; Yuan, Y. J. Evaluation of storage methods for the conversion of corn stover biomass to sugars based on steam explosion pretreatment. *Bioresour. Technol.* **2013**, *132*, 5−15.
(38) Rasmussen, H.; Sorensen, H. R.; Meyer, A. S. Formation of degradation compounds from lignocellulosic biomass in the biorefinery: sugar reaction mechanisms. *Carbohydr. Res.* **2014**, *385*, 45−57.
(39) Liu, Z. H.; Chen, H. Z. Two-step size reduction and post-washing of steam exploded corn stover improving simultaneous saccharification and fermentation for ethanol production. *Bioresour. Technol.* **2017**, *223*, 47−58.
(40) Xie, S.; Ragauskas, A. J.; Yuan, J. S. Lignin conversion: Opportunities and challenges for the integrated biorefinery. *Ind. Biotechnol.* **2016**, *12* (3), 161−167.
(41) Cherubini, F. The biorefinery concept: Using biomass instead of oil for producing energy and chemicals. *Energy Convers. Manage.* **2010**, *51* (7), 1412−1421.
(42) Liu, Z. H.; Olson, M. L.; Shinde, S.; Wang, X.; Hao, N. J.; Yoo, C. G.; Bhagia, S.; Dunlap, J. R.; Pu, Y. Q.; Kao, K. C.; Ragauskas, A. J.; Jin, M. J.; Yuan, J. S. Synergistic maximization of the carbohydrate output and lignin processability by combinatorial pretreatment. *Green Chem.* **2017**, *19* (20), 4939−4955.
(43) Bian, H. Y.; Chen, L. H.; Gleisner, R.; Dai, H. Q.; Zhu, J. Y. Producing wood-based nanomaterials by rapid fractionation of wood at 80 degrees C using a recyclable acid hydrotrope. *Green Chem.* **2017**, *19* (14), 3370−3379.
(44) Chen, L. H.; Dou, J. Z.; Ma, Q. L.; Li, N.; Wu, R. C.; Bian, H. Y.; Yelle, D. J.; Vuorinen, T.; Fu, S. Y.; Pan, X. J.; Zhu, J. Y. Rapid and near-complete dissolution of wood lignin at ≤ 80 degrees C by a recyclable acid hydrotrope. *Sci. Adv.* **2017**, *3* (9), e1701735.
(45) Tian, D.; Hu, J. G.; Chandra, R. P.; Saddler, J. N.; Lu, C. H. Valorizing recalcitrant cellulolytic enzyme lignin via lignin nano-particles fabrication in an integrated biorefinery. *ACS Sustainable Chem. Eng.* **2017**, *5* (3), 2702−2710.
(46) Frangville, C.; Rutkevicius, M.; Richter, A. P.; Velev, O. D.; Stoyanov, S. D.; Paunov, V. N. Fabrication of environmentally biodegradable lignin nanoparticles. *ChemPhysChem* **2012**, *13* (18), 4235−4243.
(47) Li, J. B.; Henriksson, G.; Gellerstedt, G. Lignin depolymerization/repolymerization and its critical role for delignification of aspen wood by steam explosion. *Bioresour. Technol.* **2007**, *98* (16), 3061−3068.
(48) Sannigrahi, P.; Ragauskas, A. J.; Miller, S. J. Effects of Two-stage dilute acid pretreatment on the structure and composition of lignin and cellulose in loblolly pine. *BioEnergy Res.* **2008**, *1* (3−4), 205−214.
(49) Cao, S. L.; Pu, Y. Q.; Studer, M.; Wyman, C.; Ragauskas, A. J. Chemical transformations of Populus trichocarpa during dilute acid pretreatment. *RSC Adv.* **2012**, *2* (29), 10925−10936.
(50) Constant, S.; Wienk, H. L. J.; Frissen, A. E.; de Peinder, P.; Boelens, R.; van Es, D. S.; Grisel, R. J. H.; Weckhuysen, B. M.; Huijgen, W. J. J.; Gosselink, R. J. A.; Bruijnincx, P. C. A. New insights into the structure and composition of technical lignins: a comparative characterisation study. *Green Chem.* **2016**, *18* (9), 2651−2665.
(51) Sun, Q. N.; Khunsupat, R.; Akato, K.; Tao, J. M.; Labbe, N.; Gallego, N. C.; Bozell, J. J.; Rials, T. G.; Tuskan, G. A.; Tschaplinski, T. J.; Naskar, A. K.; Pu, Y. Q.; Ragauskas, A. J. A study of poplar organosolv lignin after melt rheology treatment as carbon fiber precursors. *Green Chem.* **2016**, *18* (18), 5015−5024.
(52) Yoo, C. G.; Li, M.; Meng, X. Z.; Pu, Y. Q.; Ragauskas, A. J. Effects of organosolv and ammonia pretreatments on lignin properties and its inhibition for enzymatic hydrolysis. *Green Chem.* **2017**, *19* (8), 2006−2016.

575909 Tailoring the Molecular Weight of Purified Lignins for Materials Applications

Lignin is like any other polymer in that the molecular weight (MW) can have a dramatic impact on its suitability for a given application. Furthermore, polymer purity can be an equally important factor if the desired materials properties are to be achieved. Two large and growing markets have been identified for lignin use, where both MW and purity can play an important role: (1) high-performance carbon fibers for automotive applications and (2) rigid polyurethane foams for spray insulation for buildings. In both cases, we hypothesize that today's commercial lignins have too broad a molecular weight (MW) distribution and are too low in purity to give acceptable performance. Preliminary results with lignins of controlled MW and high purity are encouraging. For example, very clean, high-MW lignins isolated via ALPHA were converted into carbon fibers almost 40% stronger than any previously reported from lignin. For rigid polyurethane foams, preliminary data suggest that very clean lignins of low molecular weight and polydispersity with high hydroxyl content would be ideal as a replacement for polyols. The Aqueous Lignin Purification with Hot Agents (ALPHA) process is being investigated for controlling the MW and polydispersity, purity, and chemical functionality of lignin fractions. Hot aqueous solutions of acetic acid and acetone are each being investigated as the renewable solvents for effecting the desired separations. NMR is being used to quantify the effect of ALPHA solvent and operating conditions on the chemical functionality of the fractions being isolated.

581705 Light-weight Carbonaneous Biocomposites for Auto-parts uses

Sustainable light-weight biocomposites are the wave of the future in auto-industries for much needed fuel efficiency of the vehicles. In plastic composite uses, the key advantages of natural fibres over synthetic glass fibers and mineral fillers (talc and calcium carbonate) are their lower density, lower cost and their sustainable nature. Disadvantages of natural fibres for their use in auto-parts composites are: supply chain uncertainty, hydrophilicity and the odor and aesthetics of the finished products. Importantly, the instability of biofibre at higher temperatures limits their uses in engineering plastics (e.g. nylon, polybutylene terepththalate (PBT), etc.) based biocomposites. Traditionally, natural fibres are melt compounded with a variety of thermoplastics (e.g. polypropylene (PP), high density polyethylene (HDPE), poly(vinyl chloride) PVC), at around 180^0C. Processing temperature of engineering plastics require >250^0C. At these temperatures, natural fibres burn or degrade with burning smells. Conversion of biomass to biocarbon allows the effective use of bio-based filler in more challenging plastics, not suitable for biofibre composites. For their properties, carbonaceous biocomposites have attracted attention among academicians, scientists, and auto-makers in engineering novel biocomposite materials. A range of biocarbons can be made through pyrolysis of a variety of biomass sources, including waste biomass, municipality solid waste, and by-products from food industry and biofuel industry. In this modern era and with the circular economy concept waste valorization has attracted significant attention from several industrial sectors. This presentation will highlight the opportunities and challenges of injection moulded carbonaceous biocomposites in sustainable materials uses with focus on auto-parts.

Acknowledgements: The author would like to thank the following for their financial support: i) Ontario Ministry of Agriculture, Food and Rural Affairs (OMAFRA)/University of Guelph – Bioeconomy for Industrial Uses Research Program (Project # 030332); ii) the Natural Sciences and Engineering Research Council (NSERC), Canada Discovery Grants Project # 400320 and 401111; iii) the Ontario Research Fund, Research Excellence Program; Round-7 (ORF-RE07) from the Ontario Ministry of Research, Innovation and Science (MRIS) (Project # 052644 and 052665).

571361 Upgrading Lignin Side-Streams into Sustainable Ligniox Plasticizers and Dispersants

Introduction

Lignin is extensively studied and currently underutilized source for sustainable biochemicals, biomaterials and biofuels. Large quantities of lignin are available from kraft pulp production, even if lignin is still mainly burnt to provide process energy. Recently, also emerging biorefineries are entering the technical lignin market providing different lignin qualities, such as hydrolysis lignin from bioethanol production, and lignins from soda and organosolv pulp production.

Dispersants are widely used for suspending colloidal particles in cosmetics, paints, pharmaceuticals, oil drilling mud, cement, and ceramic applications. Besides lignosulfonates from sulfite pulp production, mainly synthetic polymers are used as dispersants in these applications. For substituting the current non-biodegrable polymers with bio-based (less expensive) materials, technical lignins have been modified in several ways, e.g. by sulfomethylation, oxidation, carboxymethylation, grafting with PEG, hydroxymethylation/propylation and sulfonation to convert them into water-soluble dispersants. Despite the encouraging results, the modification procedures seemed to be industrially unattractive and the property-performance relationships for lignin-based dispersants case dependent [1].

LigniOx technology is a simple and cost-efficient alkali-O_2 oxidation process for conversion of technical lignins for concrete plasticization, i.e. cement dispersing, or versatile dispersants [2;3]. The LigniOx process can be integrated into the biorefineries or operated in stand-alone units (e.g. by chemical industry). The oxidation is active toward phenolic hydroxyls and introduces acidic groups into the lignin polymer, while retaining its polymeric structure. Depending on the oxidation conditions, especially pH, the negative charge and molar mass of lignin can be adjusted in a controlled way. The previous studies on mortar and concrete plasticization have demonstrated that the soda and kraft LigniOx lignins have great potential to compete with lignosulfonates in their traditional markets, and even with synthetic products. LigniOx lignins have also shown high dispersing performance of several inorganic pigments, such as TiO_2, color pigment, and $CaCO_3$, used in paints, for instance [4;5]. Currently, the oxidation technology is optimized for several different lignin raw materials primarily to produce concrete plasticizers [6;7].

This study presents the very recent results of applying kraft, organosolv, and hydrolysis lignin based LigniOx lignins as plasticizers in cement mortar. In addition, the dispersing performance of LigniOx lignins in special carbon black, used e.g. in paints, coatings, inks, textile printing pastes, and aqueous graphite dispersions, is demonstrated. The potential environmental impacts of the LigniOx plasticizer production in kraft pulp mill were also studied and compared to benchmark synthetic superplasticizer from a life cycle assessment (LCA) perspective.

Experimental

Industrial lignin raw materials provided by Metsä Fibre, CIMV and St1 were used. Oxidations of different lignin raw materials were performed under the range of conditions reported in [2;3].

The plasticizing performance of oxidized lignin solutions was evaluated in mortar prepared using a standard sand mixture, Portland cement (CEM I 52.5N), and water. Plasticizer dosages 0.3 - 0.6 wt% (active matter) of cement were used. Few drops of de-airing agent were used in all tests. The fluidity of the mortar was measured using Haegermann flow table. Commercial polycarboxylate ether and sulfonated naphthalene based superplasticizers and lignosulfonate based plasticizers were used as references.

For the LCA, only the production phases of the kraft lignin based LigniOx plasticizer and the synthetic concrete superplasticizer admixture [9] were analyzed, taking out the transport, usage phases and end-of-life. Production of LigniOx plasticizer was conducted integrated to a modern kraft pulp mill [8]. The production of 1 kg of concrete plasticizer was considered as a functional unit. Five potential impact categories were assessed: global warming, acidification, eutrophication, abiotic depletion, and ozone layer depletion. The method used was CML-IA baseline.

The dispersing performance of the oxidized kraft, organosolv and hydrolysis lignin based samples was evaluated in Special Carbon Black (Orion Engineered Carbons). The dispersion prepared using high shear mixing contained 10 wt% of pigment in water. Dispersant dosages 0.25 - 2.5 wt% (active matter) of pigment were used. The viscosity of the pastes was measured as a function of shear rate at 25°C. Commercial polyacrylic acid and lignosulfonate based dispersants were used as references.

Results and discussion

Strong dispersion capabilities of the LigniOx solutions were verified in cementitious mortar. Kraft and organosolv lignin based LigniOx solutions provided equal or even better plasticization performance compared to the commercial synthetic references. The performance of hydrolysis lignin based LigniOx was a bit lower than synthetic references. Optimization of process conditions for hydrolysis lignin continues. For developing a LigniOx based concrete plasticizer, the next phases will include product formulation, verification of the plasticizing performance in fresh concrete and determination of early and matured strength of the hardened product. Field-testing will follow the laboratory scale investigations.

According to the LCA, the production of kraft lignin based LigniOx plasticizer presented the best environmental performance, showing 50-60% lower impacts in eutrophication and acidification categories, around 70-80% lower in global warming and abiotic depletion indicators and about 50% lower impacts in ozone layer depletion indicator. The main contributors to the total environmental impacts of LigniOx plasticizer produced integrated in kraft pulp mill were the use of sodium hydroxide, and chlorine dioxide (due to allocation of part of bleaching chemicals to LigniOx product).

All the tested LigniOx lignin solutions showed high dispersing performance in special carbon black when using dosing of 2.5%. The viscosity values were measured just after preparing the dispersion and after 7d. There was no changes seen after the storage period. Better dispersing performance of LigniOx lignins over commercial polyacrylic acid reference (2.5%) was supported also by microscopy and particle size analyses. Differences in the performance between the tested LigniOx samples, varying in terms of molar mass, were seen when decreasing dosing. Kraft lignin based sample (Mw of 7 kDa) was still showing high performance, while the other LigniOx samples as well as the lignosulfonate or polyacrylic acid reference could not retain the viscosity as low. The results highlight that the oxidation conditions must be optimized according to the specific end-uses to provide high-performance LigniOx products for each lignin type. Regarding the new end-uses of LigniOx lignin, performance in the actual formulated end-product (paints, coatings or inks), and the co-effect with all the other ingredients should be demonstrated.

Conclusions

LigniOx oxidation technology converts various industrial lignin raw materials to high-performing plasticizers and versatile dispersant. In addition to kraft lignin, it was shown that LigniOx solutions based on organosolv and hydrolysis lignin have high dispersing performance in cementitious material. The results of life cycle assessment of the LigniOx production integrated to a kraft pulp mill indicated better environmental performance when comparing to production of commercial synthetic concrete plasticizer admixture. LigniOx solutions also efficiently dispersed special carbon black in water, which further opens new application possibilities for oxidized lignins. LigniOx lignins are seen as a promising, bio-based and environmentally friendly alternative for several applications areas with significant market volume.

Acknowledgements

Bio Based Industries Joint Undertaking (BBI JU) under the European Union's Horizon 2020 Research and Innovation program is acknowledged for funding the project LigniOx (Lignin oxidation technology for versatile lignin dispersants) under grant agreement No 745246.

References

1. Chen, J., Kazzaz, A.E., Mazandarani, N.A., Feizi, Z.H., Fatehi, P. 2018. Review: Production of Flocculants, Adsorbents, and dispersants from lignin. Molecules 23, 868-893, doi:10.3390/molecules23040868.
2. Kalliola, A., Vehmas, T., Liitiä, T., Tamminen, T. 2015. Alkali-O2 oxidized lignin – A bio-based concrete plasticizer. Ind Crops and Prod. 74, 150-157.
3. Use of oxidised lignin as a dispersant FI20135986, US9676667, ZL201480054352.3., WO2015/049424.
4. Kalliola, A., Vehmas, T., Liitiä, T., Tamminen, T. 2017. Concrete Plasticizers and Versatile Dispersants from LigniOx lignins. Nordic Wood Biorefinery Conference, Stockholm, Sweden, March 28-30, 2017, Proceedings.
5. Alkali-O2 oxidized lignin as dispersant WO2017/077198 A1.
6. http://www.ligniox.eu/.
7. Kalliola, A., Vehmas, T., Liitiä, T., Tamminen, T. 2018. LigniOx lignins – High performance concrete plasticizers and versatile dispersants. Nordic Wood Biorefinery Conference, Helsinki, Finland, October 23-25, 2018, Proceedings.
8. Kangas, P., Kaijaluoto, S. and Määttänen, M. 2014. Evaluation of future pulp mill concepts - Reference model of a modern Nordic kraft pulp mill, NPPRJ 29(4), 620–634.
9. European Federation of Concrete Admixtures Associations Ltd. (EFCA). 2006. ENVIRONMENTAL PRODUCT.

564443 Synthesis of Acrylated Thermoplastic Polymers from Crude Lignin Bio-Oil: Effect of Intrinsic Properties of Crude Bio-Oil and Synthesis Conditions

In the present study, renewable thermoplastic polymers were produced from crude lignin bio-oil using a controlled radical polymerization technique. The bio-oil, produced from red oak lignin by pyrolysis, consists of 4% of phenolic monomers and over 95% of aromatic oligomers. The Mw and Mn of the lignin bio-oil were 653 g/mol and 388 g/mol, respectively, and there were 5.6 moles of -OH groups per mole of lignin bio-oil. The bio-oil composed of mostly multi-hydroxyl group compounds was first functionalized using a combination of partial acrylation and acetylation prior to the subsequence Reversible Addition Fragmentation Chain Transfer (RAFT) polymerization in order to control the number of polymerizable site per molecule while eliminating the radical inhibitory effect of the reside phenolic -OHs. In the present study, the effects of experimental parameters on the polymer synthesis were further studied in detail. It was found that the molecular weight of the arylated polymer and its yield both increase as the degree of the acrylation or the AIBN concentration increase. The results also showed that at least 3wt% of AIBN and 1 unit of acrylation (i.e., 1mole vinyl C=C per mole of bio-oil) are required to obtain reasonably high molecule weights and yields of the polymers. In this study, the polymers with yields up to 68% and Mw of 22kg/mol were obtained upon reacting at 90 for 4 hours. Based on the EPR analysis, it was found that AIBN is partly consumed by the intrinsic free radicals in lignin bio-oil through radical termination reactions. It was also found that the presence of a chain transfer agent can help to reduce crosslinking, therefore preventing gelation during the polymer synthesis. In this study, all resulting polymers exhibited the decomposition temperature higher than 215°C, and the temperature for maximum decomposition rate at about 350 °C. The solid residues of the polymers upon pyrolysis at 1000 °C were about 35-36%, showing their high heat tolerance. While T_gs of the polymers could reach up to 130 °C, there was a linear relationship between the M_ns of the polymers and their corresponding T_gs. The rheological results showed that the polymers synthesized using lower AIBN concentrations and lower degrees of acrylation exhibited overall reduced complex viscosity and improved meltability.

580892 Desulfurization and Coking of Hydrocarbon Fractions from Biomass Pyrolysis Oils

Fast pyrolysis is a thermochemical process for converting biomass into a crude liquid product (bio-oil), which can be hydrotreated into fuel-grade hydrocarbons. Previously, we demonstrated that advanced pyrolysis processes (e.g. tail-gas reactive pyrolysis (TGRP), catalytic fast pyrolysis) produce bio-oils that are partially deoxygenated. Subsequently, these oils undergo facile separations processes like distillation and extraction, producing a hydrocarbon-rich fraction and a phenolic-rich fraction.

Based on the hydrocarbon fraction of bio-oil, we processed the hydrocarbon fraction to produce fuel-grade hydrocarbons and calcined coke. We used bio-oils sourced from switchgrass and guayule bagasse feedstocks, the latter of which contains elevated levels of sulfur (400 – 500 ppm) due to biomass processing. Distilled fractions of bio-oil hydrocarbons were found to contain sulfur ranging from 400 – 700 ppm. Sulfur characterization revealed the sulfur groups to be labile, so Raney Nickel was used for hydrogenation, which reduced sulfur content to < 15 ppm. When co-processed with petroleum coke, the heaviest hydrocarbon fraction produces calcined coke with both a reduced metals concentration and sufficient crystallinity for aluminum smelting applications (~22 – 25 A).

587236 Effect of Lignin Hydrophobicity and Zeta Potential on Enzymatic Hydrolysis

Residual lignin plays an inhibitory role in the enzymatic hydrolysis of cellulosic biomass. In this study, we examined the structural changes in isolated lignins from hot water hydrothermal pretreatment and organosolv pretreatment and their inhibitory effects on the enzymatic hydrolysis of Avicel. The functional groups of the isolated lignins were determined using quantitative ^{13}C, 2D HSQC and ^{31}P NMR. The increase in hot water pretreatment severity significantly increased the condensed and non-condensed syringyl and guaiacyl OH group content in the isolated lignins, but decreased the aliphatic OH, p-hydroxybenzoate OH and carboxylic OH group content. A compelling adverse association ($r^2 = 0.998$) was observed between the condensed syringyl and guaiacyl phenolic OH group content in lignins and their inhibitory effects on enzymatic hydrolysis. Langmuir adsorption isotherms showed that a higher pretreatment severity resulted in a higher binding ability between the isolated lignins and the cellulase enzymes, which led to more non-productive binding. It is hypothesized that condensed syringyl and guaiacyl phenolic units are mainly responsible for the inhibitory effect of lignin on enzymatic hydrolysis, in which the condensed aromatic rings enhance the hydrophobic interactions and the phenolic OH group boosts the hydrogen bonding. The combination of hydrophobic interactions and hydrogen bonding can further intensify undesirable non-productive binding. In addition, we observed that organosolv lignins with high zeta potentials enhanced enzymatic hydrolysis.

576138 Preparation of Nanofibrillated Biomass (NFB) for Thin Films and Composites

Nanofibrillated cellulose has been used for making composites, nanopapers and thin films. However, such materials have not been constructed using native lignocellulosic biomass. Using biomass instead of cellulose allows complete utilization of plant material and may reduce chemical and energy consumption. The process for making nanofibrillated biomass (NFB) and its use in composites and thin films is described. Fibrillation of poplar hardwood was carried out using a super mass collider for producing nanofibrillated biomass. Whole biomass nanopapers were constructed by simply drying the suspensions under ambient conditions and without the use of any organic solvents. Contact angle, tensile testing, water vapor transmission rate, water retention value, transmission electron microscopy and scanning electron microscopy were carried out on the biomass nanopapers. Fiber dimension was a critical factor in making flexible biomass nanopaper. Unlike wood flour, incorporation of up to 20% NFB into polylactic acid did not cause a major loss of tensile properties of the composites. Overall, this fibrillation process allows size reduction of wood to nanoscale dimensions, which may allow production of better lignocellulosic materials.

578441 Tailoring Lignin Chemistry to Manufacture Quality Material for Sustainable and Viable Biorefienry

The success of a modern biorefinery heavily depends on the availability of value-added product streams from lignin. Nevertheless, the quality and yield of lignin-derived materials heavily depends on the fundamental understanding of lignin chemistry. In particular, there is an imperative need to define the relationship between lignin chemical characteristics and performance of materials like carbon fiber, asphalt binder, and nanoparticles. In the past several years, our multidisciplinary team has significantly advanced the understanding of how to tailor lignin chemistry to improve the performance of lignin-based materials. For carbon fiber, we have revealed that carbon fiber mechanical and conductive performance heavily depends on the molecular weight, uniformity, chemical linkages, and functional group profile of lignin. The chemical characteristics determines the miscibility of lignin with guest polymer and thus impacts the crystallite content and size, resulting in carbon fibers with various performance. For nanoparticles, we have shown that the functional groups, chemical linkages, molecular weight, S/G ratio, and lignin condensation all could impact the hydrogen bond networks and electron double layers, which could in turn define the size, uniformity, and stability of lignin nanoparticles. For asphalt binder modifiers, the functional groups and molecular weight also impact the interaction with asphaltene, thus defining the high temperature and low temperature performance of asphalt binder. Overall, lignin chemistry is crucial for the performance of various high value products. The fundamental understanding transformed the biorefinery design, where new pretreatment, fractionation, and feedstock development strategies are being developed to tailor lignin chemistry toward best-performing lignin-based products.

575788 Production of Hydrocarbons from Fast Hydropyrolysis of Lignin at Atmospheric Pressure Promoted By Nickel inside Carbon-Nanotube Reactors

We present results obtained with a novel catalyst which promotes production of liquid hydrocarbons from lignin in a single stage at atmospheric pressure. The conversion can be accomplished by catalytic hydropyrolysis of lignin over Ni-loaded carbon nanotubes (CNTs). We report that the position of the nickel catalyst at the carbon nanotubes (inside or outside CNTs) play a very important role in the process. The nickel inside the CNTs led to confinement effects that significantly improved the yields of hydrocarbons compared to results obtained with Ni outside the CNTs. We also evaluated the effect of the nickel loading and the catalyst-to-lignin ratio. The total yield increased with the catalyst-to-lignin ratio and Ni loading.

The condition that maximized the yield (12.53 %) was the 20:1 catalyst-to-lignin ratio with 15 wt % Ni loading. The most abundant aromatic compounds were benzene, toluene, xylene, and indene. The most abundant aliphatics compounds were alkanes, alkene, cycloalkanes and alkanols. Aromatics and aliphatic compounds are important components of jet fuels. Given the well-known difficulties in thermal processing of lignin, the results of this work obtained at atmospheric pressure provide a promising route for the direct conversion of lignin into drop-in transportation fuels.

576579 Kinetic Study of Tire Carbon Black and Biomass Soot Steam Activation Used for the Removal of Phenol and Chlorine from Drinking Water

This study aims to demonstrate a novel method for removing toxic chemicals using soot produced from wood and herbaceous biomass pyrolyzed in a drop tube reactor and tire pyrolytic carbon black. In order to improve the economic performance and reliability of entrained flow gasification, biomass soot has the potential to be captured, activated and further used as a sorbent to remove a broad range of pollutants from wastewater. In recent years, the disposal of waste tires has become an important issue. The liquid hydrocarbons and solid char residue from thermal conversion of waste tires have potential to be used as environmentally benign fuel and activated carbon. In the present work, the influence of ash content, nanostructure, particle size, and porosity on the filter efficiency of steam activated carbon materials was studied. It has been shown for the first time that steam activated soot and carbon black can remove phenol and chloride with the filter efficiencies as high as 95 %. The correlation of the filter efficiency to material properties showed that the presence of alkali and steam activation time were the key parameters affecting filter efficiencies. Kinetic parameters for the steam activation of tire pyrolytic carbon black and biomass soot have been determined by thermogravimetric analysis using steam as an activation agent. The feedstock for the soot production, activation temperatures and residence time were varied. The results showed that the rate of steam activation is 10 times faster than the activation at lower temperature. This study demonstrated that the steam activation process is suitable to prepare activated soot with the high surface area at the short residence time and at temperatures below 800°C. Reaction evolution has successfully been modelled, including the maximum reaction rate obtained in experimental results. Moreover, intrinsic kinetic parameters of the activation reaction order, activation energy and pre-exponential factor – were calculated.

576691 The Effect of Wood Composition and Supercritical CO_2 Extraction on the Charcoal Production

This work demonstrated that the coupling of supercritical carbon dioxide extraction with slow pyrolysis is effective to remove over half of extractives from low quality wood and to generate biochar from remaining solid wood fractions. The high yields of extractives from supercritical carbon dioxide extraction illustrates the potential utilizing of low quality wood as an alternative feedstock for the sustainable production of value-added chemicals. Results showed that supercritical carbon dioxide extraction has neither a strong impact on the physical properties of original wood nor on the yield of solid biochar. These results are promising as they show that the biochar obtained for this renewable feedstock could be used as an alternative to fossil-based coke in applications including ferroalloy industries. Moreover, the heat treatment temperature and supercritical carbon dioxide extraction had a significant impact on the tar yields, leading to the increase in naphthalene, polycyclic aromatic hydrocarbons, aromatic and phenolic fractions with the greater temperature. The differences in gasification reactivity and dielectric properties of solid biochars, composition and yields of liquid products of non-treated pinewood and extracted wood fraction emphasize the impact of supercritical carbon dioxide extraction on the pyrolysis process.

576979 Lignin Hydrogenolysis with in-Situ Glycerol Aqueous Phase Reforming

Lignin hydrogenolysis was conducted with glycerol aqueous phase reforming (APR) in the in-situ reaction. Given that glycerol, a by-product of the biodiesel process, produces hydrogen and lignin, a lignocellulosic constituent, produces renewable phenolics, a substitution for petroleum-based chemical compounds can be obtained. In order to understand the mechanistic reaction of the glycerol APR reaction and the in-situ reaction, the initial study was conducted with phenol as a substitute for the lignin model compounds. Depending on local hydrogen availability, glycerol APR was found to proceed in parallel to form ethylene glycol or 1,2 propylene glycol. Hydrogenation of phenol was found to scavenge hydrogen produced from APR leading to the suppression of the undesired reactions of glycerol APR. Based on the fundamental study on the reaction mechanism of in-situ hydrogenation and aqueous phase reforming, various lignins and bio-oil have been depolymerized and hydrogenated. Comparative results between various technical lignins and bio-oils will be presented in terms of product distribution and monomeric yield, as well as interesting phenomena during the reaction.

577215 Synthesis of Photocatalytic Hydrochars and Application of Emerging Contaminants Degradation

Hydrothermal carbonization of waste biomass has led to emergence of hydrochars – a porous carbon material – for environmental applications. We aim to improve hydrochar's application in the field of water and wastewater treatment by creating a novel photocatalytic nanocomposite from deposition of homogeneously dispersed TiO_2 nanoparticles on hydrochar surface. This conjugated nanocomposite combines with important advantage of an excellent adsorbent i.e., hydrochar for emerging organic pollutant adsorption with the ability to degrade these contaminants by photocatalytic activity of TiO_2 nanoparticles. In this study, we prepared TiO_2 nanoparticles *in situ* during the hydrothermal carbonization of food waste. We studied the effect of temperature and precursor ratios (biomass:TiO_2 precursor) on the physicochemical properties of the synthesized hydrochar-TiO_2 nanocomposite. This was done using scanning electron microscopy (SEM) for physical morphology, electron dispersive spectroscopy (EDS) for elemental composition, and thermogravimetric analysis (TGA) for metal content. We studied the degradation of ibuprofen and carbamazepine – the two well-known pharmaceuticals present in wastewater. We found that the nanocomposites degrade these compounds much faster than the TiO_2 nanoparticles themselves under sunlight activation. The bandgap measurement shows that TiO_2 achieves visible light activity when conjugated with hydrochar and that contributes to the enhanced performance in the enhancement of contaminant degradation.

577402 Analysis of Pyrolysis Oil Model Compounds Using *in-Situ* Raman

In recent decades, the world has seen an exponential increase in the demand for fossil-based energy sources. This increase in the demand is leading to an exhaustion of the petroleum reserves and increase in the environmental pollution. Therefore, attention is being given to alternative, carbon-neutral renewable energy sources like biofuels. The carbon-neutrality and lower production costs make the production of biofuel from lignocellulosic biomass an alluring alternative to conventional fossil fuels. Biofuels are generally produced from biomass via pyrolysis which yields pyrolysis oil that is high in oxygenated compound. The pyrolysis oil is then upgraded to biofuel through hydrodeoxygenation (HDO). Analysis of the HDO reactions for conversion and kinetic parameters often requires gas chromatography (GC) with time-consuming material preparation steps prior to the GC analysis. An alternate, fast analytical method would be beneficial to both researchers and producers alike.

Therefore, in this research, we propose the use of *in-situ* Raman Spectroscopy which can yield us time-resolved information about molecular structure, intermediates, mechanisms, and rate of reaction. The HDO reaction was simulated by analyzing multiple samples of varying concentrations of the reactants, and their corresponding HDO products. The obtained Raman spectra was then deconvoluted using Gaussian Deconvolution algorithm, in order to identify, and differentiate between different peaks. The height of their respective characteristic stretching modes were then measured to determine the conversion of the model compound. The carbonyl ($\nu_{C=O}$) stretching mode in hydroxyacetone, for instance, was found at 1,721 cm^{-1}, (ν_{COCOC}) sym. stretching mode of the acetal group found in glycolaldehyde diethyl acetal was observed at 1,140 cm^{-1} and the phenol ($\nu_{C\text{-}O}$) stretching mode of 4-n-propylphenol was obtained at 1,204 cm^{-1}, respectively. A calibration curve was then obtained from the corresponding peak height versus concentration. Insofar, *in-situ* Raman could be used to measure the conversion of the hydrodeoxygenation of pyrolysis oil compounds.

559892 Pyrolysis and Steam Gasification Characteristics of Mangroves Under Rapid Heating Conditions

Mangroves are one of the important resources from the viewpoint of biodiversity. On the other hand, economic development in tropical countries is causing a decline in mangroves. It is thought that positive utilization of mangroves as well as its conservation are necessary in order to pursue both sustainable development and environmental preservation. The objective of this study is to clarify the characteristics of pyrolysis and steam gasification when using mangroves. The behaviors of pyrolysis and steam gasification of them are investigated and the effect of tree species and gasification conditions, such as temperature and atmosphere, on thermal decomposition characteristics are discussed. In this connection, the role of minerals in woody biomass was discussed, because it is presumed that minerals in woody biomass such as alkali and alkaline earth metals act as catalysts for steam gasification of woody biomass.
Three kinds of mangroves (Rhizophora mucronata, Bnguiera cylindrical and Avicennia marina) and Japanese cedar, Japanese cypress, eucalyptus were used as raw materials. The biomass sample was milled and sieved into 0.5~1.0 mm. After that, 1.0±0.0009 g of the sample was weighed out and dried at 110 °C for 24 h. When experiments of the gasification are performed using demineralized sample, these were prepared by acid treatment described as follows. The biomass sample was soaked in HCl (0.58 mol/L) and stirred at room temperature for 1h. Thereafter, these were filtered, thoroughly washed and dried at 110 °C for 24 h. Pyrolysis and gasification were carried out in the batch type tubular reactor. The inside of the reactor was under inert atmosphere with argon gas (1 L/min). The heating temperature was 800 °C. The biomass sample filled in ceramic board is supplied to the heating part in reactor tube by a manual slide method, after that the thermal decomposition reaction occurs immediately. Steam gasification was carried out by adding water (0.2 mL/min) into the reactor with a syringe pump.
The current main results indicate that mangroves have high reactivity under steam gasification conditions in comparison with the other woody biomass. The produced gas yields of the mangroves under inert atmosphere were 49.9~54.0 wt%, while the gas yields of the other woody biomass were 53.3~61.7 wt%. At the same time, the gas yields of the mangroves under steam atmosphere were 68.2~81.5 wt%, while the gas yields of the other woody biomass were 65.8~73.7 wt%. From these results, the gas yields by steam gasification using the mangroves were 1.3~1.6 times greater than those under inert atmosphere, while those using the other woody biomass were 1.2~1.3 times. The concentration of both H2 and CO2 also increased with increase in conversion ratio under steam gasification conditions. This result leads us to presume that the steam reforming and water gas shift reaction are occurring remarkably in the mangroves. The catalytic function due to minerals in biomass ash is one of the reasons why the reaction between biomass and water vapor is promoted during the gasification process. Therefore, the catalytic function of the minerals was investigated, based on the gasification behavior using demineralized biomass sample as raw materials. The gas yields by steam gasification using the demineralized mangroves were almost similar to those under inert atmosphere and those were 0.96~1.06 times. The results using the other demineralized woody biomass indicated almost the same tendency. From these results, it was found that the demineralized biomass sample has poor reactivity under steam gasification conditions. It is supposed that the minerals in biomass play the role of catalysts to promote steam gasification reaction. From the above results, it was shown that the mangroves can be applied to the gasification process as well as the other woody biomass.

566195 Super Activated Hydrochar for Hydrogen Storage

The world is moving towards clearer energy. Hydrogen (H_2) is a source of high gravimetric energy as well as does not produce any carbon dioxide (CO_2) emission. The storage of H_2 is a big concern in today's world which require high pressure (around 700 bar), low temperature (around -196 °C), and expensive storage material. As a result, the use of H_2 energy is not so popular throughout the world. Various efforts have been undertaken for storing H_2 both chemically and physically. Physical storage has proven more economically viable, while considering the recovery of H_2. Along with the adsorption parameters (e.g., high microporous surface area, large pore volume, etc.), the adsorption sites and the interaction between adsorption sites with hydrogen are vital for storing large volume of hydrogen.

In the last few years, researchers have reported that the surface area and pore types of materials are the key factors for storing H_2. Recently, it has been observed that the presence of oxygen functional groups also plays an important role in uptake H_2. However, the effect of the quantity and type of oxygen functional groups is missing in the literature. In addition to this gap, as per our knowledge, no work has been done on the stability of storage materials as they are experienced with low temperature and high pressure.

Therefore, the objective of this study was to investigate the influence of surface area, pore type, presence of oxygen functional groups on H_2 uptake. Hydrochar was produced from cellulose at 260 °C. Produced hydrochars were chemically activated by KOH at 3 different temperature (600, 700, and 800 °C). BET analysis was conducted to determine the surface area and pore type of the activated chars. In addition, Boehm titration was directed to calculate the oxygen containing functional groups. Finally, Intelligent Gravimetric Analyzer was used to measure H_2 uptake. Analysis showed that, activated char with rich oxygen containing functional groups can store more H_2 compare to the standard materials such as metal organic frameworks (MOFs), covalent organic frameworks (COFs), etc.

566349 Investigations of Properties of Carbon Quantum Dots Obtained from Hydrothermally Carbonized (HTC) Process Liquids

Carbon quantum dots (CQDs) are spherical nanomaterials with a particles size range of 2 to 10 nm. CQDs exhibit excitation-wavelength dependent photoluminescence behavior and they are considered as cost effective and nontoxic alternative to the conventional inorganic quantum dots. CQDs have wide range of applications such as medical diagnostics, bio-imaging, biosensors, solar cells, photo catalysis, and others. In this study fluorescent carbon quantum dots (CQDs) were extracted from hydrothermal carbonization (HTC) process liquids prepared from hydroxymethyl furfural (HMF), furfural, and cellulose. HTC experiments were performed at 200, 220, and 240 °C for a reaction time of 30 min. Dark brown and yellowish HTC process liquids were centrifuged at 10,000 rpm for 10 minutes and then fluorescent CQDs containing solutions were separated. CQDs containing solutions were filtered with syringe filter and finally, crystalline CQDs were obtained from vacuum distillation and freeze drying. Different properties of CQDs were observed by TEM, FT-IR, XPS, UV-Vis for different feedstocks as well as HTC conditions.

567993 Wet Biomass Waste Valorization through Hydrothermal Liquefaction and Integrated Post-Treatment Measures

Wet biomass wastes such as food waste and dairy waste have very high water content over 90%. They can cause severe environmental and public health issues if left untreated including surface/groundwater pollution, water eutrophication, greenhouse gas emissions, etc. Traditional thermochemical treatments (e.g., pyrolysis or combustion) are energy inefficient for wet biomass wastes due to the higher vaporization enthalpy requirement. Hydrothermal liquefaction (HTL) hydrolyzes organics contained in wet biomass in an aqueous media at temperature ranging from 250 to 350 °C and pressures from 3 to 30 MPa for residence times of 5 to 60 minutes to produce bio-crude oil, solid hydrochar, and water-soluble chemicals in varying amounts depending on feed composition and operating conditions. Bio-crude oil can be further upgraded and used for transportation fuel. Hydrochar can act as a soil amendment due to its high nutrient (N and P) content. Water-soluble chemicals include short-chain carboxylic acids and sugar monomers that can be recovered through post-treatment measures such as anaerobic digestion (AD) to produce biogas for electricity generation or recovered as phosphorous-rich and nitrogen-rich nutrient compounds using membrane distillation (MD).

Currently, most researchers employ batch methods for HTL processing of wet biomass wastes where the feedstock(s) are loaded into the reactor and then heated to the target temperature. The heating time, depending on the target temperature and reactor size, can vary from several minutes up to an hour. However, the results obtained from batch experiments often are not quantitatively reproducible. In our studies, we have found that batch results differ significantly from continuous HTL operation on the industrial scale where the feedstock is injected into the reactor maintained at the target temperature. We examine how different experimental methods (i.e. batch vs. feed injection) affect the product distribution for a variety of different feedstocks including model compounds and mixtures representing typical wet biomass wastes. It explores methods to optimize the target product yield (i.e. bio-crude oil) in terms of mixture composition in order to maximize synergistic effects among different compounds.

In the presentation, post-treatment measures for the aqueous phase including AD and MD will also be briefly outlined to illustrate the benefits of combining HTL with other treatment methods. Along those lines, system analysis studies conducted in our group of applications of combined AD and HTL processing of dairy and food wastes will be reviewed.

572333 Scale up and Product Validation of Biomass Derived Sugar to Acrylonitrile Process

Ninety percent of the world's carbon fiber production utilizes acrylonitrile as a raw material, growing at 9 to 14 percent per year. Due to the high growth rate of carbon fiber production, any reduction in GHG originating from Acrylonitrile production will be highly impactful. Conventionally, acrylonitrile is produced from propylene ammoxidation. However, the current method makes use of an alternative approach using sugars obtained from biomass feedstock. Southern Research in partnership with DOE/EERE is developing a method to provide renewable acrylonitrile produced from biomass-derived second generation sugars.

An innovative, thermo-catalytic process has been developed that converts second generation sugars obtained from biomass to acrylonitrile. The proposed process takes place in three reaction steps starting with biomass derived sugars. In the first reaction step sugars obtained from biomass hydrolysis are hydrocracked to obtain high purity glycerol using a proprietary catalyst. In the second step, a diluted version of glycerol is used for dehydration reaction in which glycerol is converted to acrolein which is then subjected to ammoxidation reaction in the third step to obtain acrylonitrile.

Our previous AIChE presentations have focused on the laboratory scale study and results. Following successful laboratory scale demonstration, the project has moved to bench scale where the process has been scaled up to 1000x. Catalysts have been synthesized in near kg batches and the productivity from each reaction step elevated to almost a kg/hr. In addition to studying catalyst, scale up, life for over 500 hrs. of continuous operation, the collected and purified acrylonitrile (BioACN) from the final step has been validated with the help of the commercial carbon fiber manufacturing partner, Cytec-Solvay. In this part, Cytec-Solvay conducts a systematic study to understand the impact of the extent of each possible impurity present in the BioACN to determine the maximum acceptable amount and provides feedback on final BioACN purification.

This presentation will provide an update of the ongoing bench scale study. The results from long term catalytic tests at this scale will be presented along with how some critical scale up challenges that have been addressed to improve performance. On the product side, the final product validation step will be discussed. Finally an updated comprehensive TEA/LCA study will be presented to highlight process merit and the key drivers.

572486 Fate of Inorganics during Hydrothermal Carbonization of High Ash Fraction of Air Classified Biomass

Feedstock preparation and transportation are two major upstream challenges for any biorefinery. Biomass is often wet (>60% moisture), has low bulk density (60-300 kg m^{-3}) and contains high ash (1-25 wt%). High moisture and low density significantly limit the transportation and storage of biomass, while high inorganics, especially alkali and alkaline earth metals and phosphorus (AAEMs+P), promote scaling and fouling on heat transfer surfaces, and often catalyze to produce undesirable products. It has been shown that air classification can economically remove fractions of forest residues that have high ash content; however, however, a significant fraction of organics is removed as well.

The goal of this study was to reduce the AAEMs+P from the high ash fraction of air classified forest residues (AC-FR) using Hydrothermal Carbonization (HTC) because HTC has been proven effective for leaching monovalent and divalent cations as well as halides. Therefore, high ash fractions of AC-FR was subjected to HTC at temperatures of 180-260 °C. Solid hydrochars were characterized with induced coupled plasma (ICP), ultra violet visible spectroscopy (UVVIS), ultimate analysis, proximate analysis, and scanning electron microscopy-electron dispersive X-ray spectroscopy (SEM-EDS). Meanwhile, HTC process liquids were characterized ICP and UVVIS in order to perform mass balance of AAEMs+P. The results showed that AAEMs have leached from solid to liquid and a higher HTC temperature increases the leaching. It was also observed that higher HTC temperature reduces elemental carbon and oxygen from solid. Therefore, the overall ash content of hydrochar remains similar at higher HTC temperatures, despite of an overall reduction of AAEMs from solid.

575676 High-Quality Biocrude Production from Reactive Catalytic Fast Pyrolysis

High-Quality Biocrude Production from Reactive Catalytic Fast Pyrolysis

Joseph Weiner, Phillip Cross, Elliot Reid, Ofei D. Mante, David C. Dayton

RTI International, Research Triangle Park, NC 27709 USA.

Abstract

Reactive catalytic fast pyrolysis (RCFP) is a direct biomass liquefaction process that yields a low-oxygen content hydrocarbon-rich biocrude which can be hydrotreated to produce advanced biofuel (gasoline and diesel). The process is operated under a reducing atmosphere, hydrogen gas, which drives further catalytic hydrodeoxygenation of the pyrolysis vapors to reduce biocrude oxygen content and inhibit char formation. The lower oxygen content biocrude is more thermally stable compared to bio-oil produced from conventional non-catalytic, thermal pyrolysis. Lower char yields also translate into higher biocrude yield, of the integrated RCFP process, from feedstock conversion to upgraded biofuels, is being evaluated by RTI to demonstrate the technical feasibility and economic viability of this advanced biofuels pathway.

An extended experimental campaign is currently being conducted to generate enough biocrude (~3 gallons) from the RCFP process for hydrotreating studies. During this campaign process conditions have been optimized and catalyst deactivation behavior has been studied in detail. Dried loblolly pine with a particle size distribution between 160-500 µm is fed into a laboratory-scale (2.5" diameter) bubbling fluidized bed reactor containing a commercially-available molybdenum-based catalyst maintained at 500°C in a large excess of hydrogen. Pyrolytic vapors are collected in a condensation train that includes a heat exchanger, a series of impingers, and an electrostatic precipitator. The biocrude is collected and analyzed using GCMS, TGA, and elemental analysis. An online micro GC is used to measure the composition of the permanent gases. Biocrude oxygen contents of less than 10% have been measured by the lab-scale system used at RTI.[1] Material and element balances for the RCFP process and detailed biocrude compositions will be presented. Insights into catalyst deactivation mechanisms and long-term catalyst performance will be also be presented. Preliminary results for RCFP biocrude upgrading will be compared to previous results presented in the literature.[2]

Keywords: bioproducts, reactive catalytic pyrolysis, hydrotreatment, biofuel

References:

K. Wang, D. Dayton, J. Peters, O. Mante, *Green Chem.*, 2017, **19**, 3243-3251.

Mante, O. D., D. C. Dayton, J. Gabrielsen, N. L. Ammitzboll, D. Barbee, S. Verdier, K. Wang

Green Chem., 2016, **18**, 6123-6135.

565464 An Electric Drive for Biorefineries: Electrochemical pH-Shift Liquid-Liquid Extraction of Succinic Acid

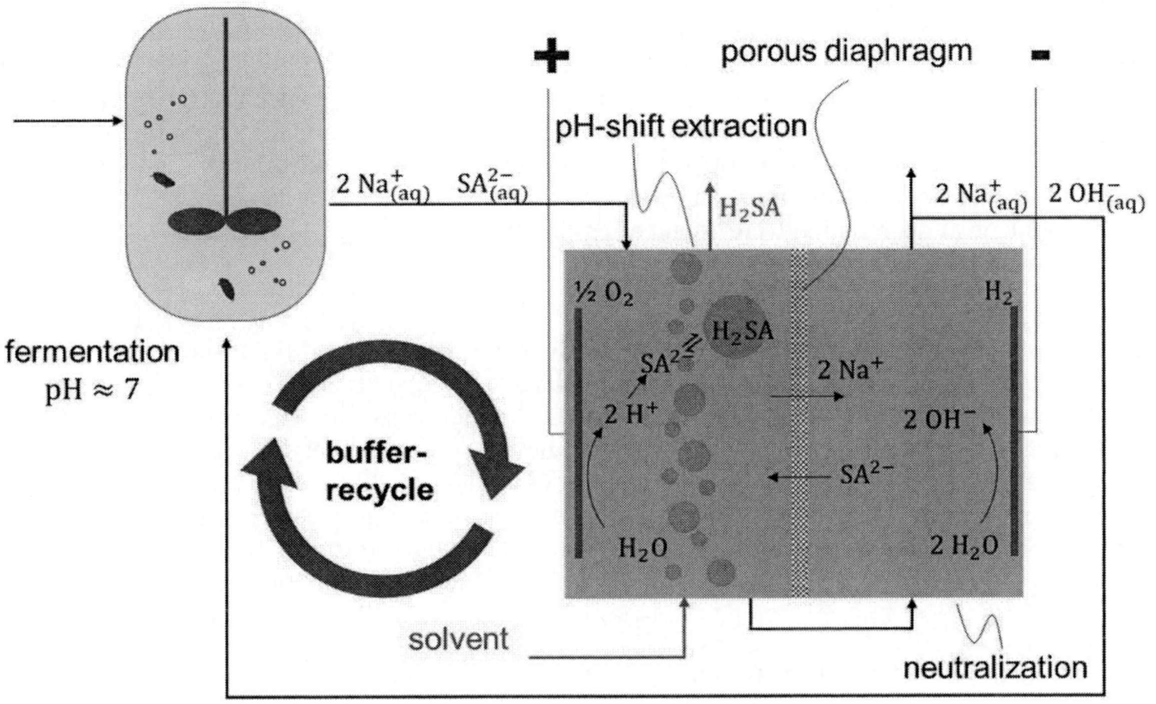

The progress in industrial biotechnology provides new bio-based routes for the sustainable production of platform chemicals.[1] Biotechnological production of succinic acid (SA) is a promising example where bio-based processes now compete with established petrochemical production routes.[2] Succinic acid is a four carbon (C4) di-carboxylic acid which makes it a versatile building block for bio-based polyester and nylon-type polymers or a precursor for solvents like 1,4-butanediol and tetrahydrofuran.[3] The development of well-performing bio-catalysts shifted the bottleneck of bio-based succinic acid production from upstream to downstream operations. In the manufacturing of carboxylic acids, separation and product purification can account for up to 40% of total production costs.[4] The high separation costs originate from a mismatch between the optimal pH for fermentation and product recovery. While most bio-catalysts work best at neutral pH, acidic pH is advantageous for product recovery.[5]

However, pH control by inorganic bases and acids comes with the drawback of producing large amounts of inorganic salt waste like gypsum.[6] Hence, new separation technologies which overcome the excessive application of inorganic acids and bases for pH adjustment are a key element for improving the competitiveness of bio-based production routes for succinic acid.

In this work we propose electrochemically driven pH-shift liquid-liquid extraction as a new method for the recovery of succinic acid from pH neutral aqueous solution. The proposed approach features the generation of spatially separated regions of acidic and alkaline pH by water electrolysis. The anodic reaction of water electrolysis produces protons, which create an acidic pH near the anode. This acidic pH shifts the aqueous acid-base equilibrium towards the protonated succinic acid, which is then extracted from the aqueous solution by an organic solvent.[5] Hydroxide ions produced at the cathode create an alkaline catholyte, which can be used for pH control in succinic acid fermentation. The water electrolysis coupled to liquid-liquid extraction creates an integrated buffer-recycle between fermentation and product separation and overcomes the need for adding inorganic acids and bases for pH-control. In contrast to electrodialysis where membrane stability and high mass transport resistance reduce separation efficiency[4] no ion selective separation of anolyte and catholyte is required in our approach.

In order to prove feasibility, we conducted batch experiments of electrochemical extraction in a 140ml setup. Thereby separation of succinic acid from aqueous solution at initial pH of 7 was achieved with tri-n-octylamine and 1-octanol as organic solvent. The change of succinic acid concentration in the aqueous phase was measured by HPLC and follows farady's law of electrolysis. Based on the concentration change, we determined a faradaic efficiency of 30-60% for separation in the non-optimized laboratory setup. The mass balance of succinic acid holds within an experimental error of 3-5%. In addition, no byproducts or indications of succinic acid degradation were found after electrolysis.

Furthermore, we developed a dynamic model of the electrochemically operated liquid-liquid extraction. A set of differential and algebraic equations (DAE) for mass balances, aqueous acid- base-equilibrium as well as mass transport kinetics of liquid-liquid extraction was implemented in MATLAB. The model was validated with the experimental results and can predict the experimentally observed change in concentration. In order to investigate its application to control the pH of succinic acid fermentation, the dynamic model for electrochemical pH-shift extraction was also implemented as s-function in the MATLAB Simulink environment and coupled to a dynamic fermentation model based on Monod-kinetic type reaction rates considering pH-Inhibition of the succinic acid producing microorganism *A. succinogenes.*[7]

Simulation results indicate that pH-management by electrochemical liquid-liquid extraction seems feasible and the simulated fermenter achieved a space time yield of 2.15 $g\ l^{-1}h^{-1}$ which is comparable to reported experimental values of 1.8-3.9 $g\ l^{-1}h^{-1}$.[8] A case study for flexible utilization of electricity was conducted to investigate the impact of intermitted electricity supply from renewable sources on the overall process performance. The results indicate that kinetics of bio-reaction are much slower than the kinetics of electrolysis and liquid-liquid extraction. Thus, a fluctuating electricity supply does not compromise the process. In a sensitivity study we set a target for faradaic efficiency, cell voltage and electricity price at which electrochemical pH-shift extraction becomes a competitive option for succinic acid downstream processing.

References

[1] J.B. McKinlay, C. Vieille, J.G. Zeikus, *Applied microbiology and biotechnology* **2007** *76 (4)*, 727. DOI: 10.1007/s00253-007-1057-y.

[2] N. Nghiem, S. Kleff, S. Schwegmann, *Fermentation* **2017** *3 (2)*, 26. DOI: 10.3390/fermentation3020026.

[3] J. Becker, A. Lange, J. Fabarius, C. Wittmann, *Current Opinion in Biotechnology* **2015** *36*, 168. DOI: 10.1016/j.copbio.2015.08.022.

[4] C.S. López-Garzón, A.J.J. Straathof, *Biotechnology advances* **2014** *32 (5)*, 873. DOI: 10.1016/j.biotechadv.2014.04.002.

[5] T. Kurzrock, D. Weuster-Botz, *Biotechnology letters* **2010** *32 (3)*, 331. DOI: 10.1007/s10529-009-0163-6.

[6] A.A. Kiss, J.-P. Lange, B. Schuur, D.W.F. Brilman, A.G.J. van der Ham, S.R.A. Kersten, *Biomass and Bioenergy* **2016** *95*, 296. DOI: 10.1016/j.biombioe.2016.05.021.

[7] S.K.C. Lin, C. Du, A. Koutinas, R. Wang, C. Webb, *Biochemical Engineering Journal* **2008** *41 (2)*, 128. DOI: 10.1016/j.bej.2008.03.013.

[8] C.D. van Heerden, W. Nicol, *Biochemical Engineering Journal* **2013** *73*, 5. DOI: 10.1016/j.bej.2013.01.015.

571724 Design and Optimization of an Acetic Acid Recovery System Via Extraction-Distillation Using an Isopropyl Acetate+Isopropanol Mixed Solvent

Azeotropic distillation incorporating liquid-liquid extraction was used for the dehydration of dilute HAc solution. A hybrid extraction-distillation process was designed and simulated for isopropyl acetate+isopropanol (IPA-01+IPA-02), and the same process using pure isopropyl acetate (100IPA-01) as a solvent was also simulated as a comparison. The effect of IPA-02 content in the solvent on the whole process was studied and the best solvent composition was determined. An optimal composition of the solvent was determined to be the mixture of 82 wt% IPA-01 and 18 wt% IPA-02 (82IPA-01/18IPA-02). The hybrid extraction-distillation process design was carried out using 82IPA-01/18IPA-02 and 100IPA-01 as solvents, respectively. The optimal design of the hybrid extraction-distillation process was achieved by the minimization of total annual cost (TAC) using Aspen plus. The result shows that the operating costs and capital costs of the process using IPA-01+IPA-02 solvent are lower than that using 100IPA-01 solvent. The optimal design of the hybrid extraction-distillation process making use of 82IPA-01/18IPA-02 as a solvent brought a reduction of the total annual cost by about 35.7% compared to that of the IPA-01 solvent.

568239 Reactive Extraction of Acetic Acid from Liquid Phase Pyrolysis Oil and Solvent Regeneration By Distillation

Biomass is also considered as an alternative to fossil resources for fuel production. To improve efficiency of biobased processes, complete utilization of biomass is essential. Therefore, also side streams are investigated for usage. Exemplarily in liquid phase pyrolysis, like the bioCRACK process, two liquid product phases are formed. The nonpolar phase, the bioCRACK oil, can directly be processed in a refinery. The polar phase, the liquid phase pyrolysis oil (LPP), has a high water fraction containing carboxylic acids and long-chain molecules. For upgrading of LPP oil to fuel grade, a hydrodeoxygenation step is necessary for removing the high oxygen content. Before hydrodeoxygenation, the carboxylic acids shall be removed by liquid-liquid extraction with the aim of generating an additional product and for increasing the catalyst stability in hydrodeoxygenation.

Carboxylic acid extraction from aqueous phase is well investigated using organophosphorus compounds, enabling distillation for solvent regeneration. Therefore, Cyanex®923, a mixture of trialkylphosphine oxides, in *n*-undecane was chosen for LPP oil extraction. To obtain sufficient separation of the acids and simultaneously minimize the loss of other organic components to the solvent phase, the concentration of the reactive extractant was optimized. The molar ratio of Cyanex®923 to acids in the LPP oil of 0.75 showed the highest selectivity towards acids of 30 %, and an acid extraction efficiency of 47 % was achieved in one equilibrium stage.

For solvent regeneration, the laden solvent was distilled at a pressure of 200 mbar in a Dean-Starck apparatus. *n*-undecane forms a heterogeneous azeotrope with water which also transfers the acids to the distillate. The aqueous distillate is collected in the product funnel, while the organic phase, the upper layer in the product funnel, is recycled to the reboiler. The acid recovery from the laden solvent to the distillate was 78 %. 89 % of water was transferred from the refined solvent to the distillate. The distillate also contained small amounts of methyl acetate, propanoic acid, butanoic acid, *n*-undecane and 1-hydroxy-2-propanone.

In a three-stage process with extraction followed by distillation for solvent regeneration, an overall acid extraction efficiency of 81 % was achieved without loss of reactive extractant to the LPP oil.

572894 Modulation of Lignin Inhibition on Lignocellulose Hydrolysis

Lignin negative impact on hydrolysis of sugarcane was reduced while maintaining efficient conversion through modulation of the liquid hot water pretreatment. A decrease in production costs require that enzyme costs be reduced. This is achieved by reducing protein loadings during the hydrolysis of lignocellulose. The addition of non-catalytic proteins has been shown to reduce lignin adsorption of cellulases and reduce the amount of the enzymes necessary. However, these proteins also incur costs albeit less than active enzymes. In this work, an alternative method to reduce lignin adsorption and protein loadings was investigated. The processes were evaluated based on relative conversion of cellulose to glucose and the improvement to it with the addition of inert protein. By controlling the pretreatment severity, through temperature and time, an optimal pretreatment condition was identified. Liquid hot water pretreatment conducted at 200°C for 20 minutes, reduced the lignin impact on final conversion to negligible. This effect was observed across different enzyme products at low protein loadings. Across literature data, a relationship between lignin content and inert protein addition can be observed. The effect of lignin adsorption is highly dependent on pretreatment method and enzyme product used. Despite this limitation, the negative impact of lignin can be predicted. This work demonstrates that optimal processing can reduce enzyme loadings towards a more efficient hydrolysis process.

581034 Modeling and Optimization of the Hydrolysis of Cellulosic Fibers by Integration of Membrane Separations and Enzyme Recycle

The conversion of waste paper fibers into sugars by hydrolysis of their cellulosic content can provide a cheap and high grade feedstock for various small and large molecule products. The hydrolysis process with enzymes is difficult though on account of the presence of mineral inhibitors in recycled pulps and also the lower accessibility of cellulose due to fiber hornification effects. Effective adjuvants such as surfactants can reduce inhibitory effects. The enzymatic hydrolysis process can be very effectively enhanced by integrating proper separation steps with the reactor and thereby effectively recycle the reactants.

The present contribution describes an effective integrated membrane bioreactor for this application and a mathematical model based on hydrolysis kinetics. The operation of the separation is optimized with the model and predictions are confirmed experimentally.

570116 Robust, High-Performance Graphene Oxide Nanofiltration and Reverse Osmosis Membranes for Black Liquor Concentration

Black liquor (BL) concentration in the kraft process is extremely energy intensive due to the use of multi-effect evaporators. Membrane-based processes can save large amounts of energy, but the development of robust (long-lived) and high-performance membranes for BL concentration has remained elusive and has been highly challenging for conventional polymeric and ceramic membranes [1]. We will discuss the ongoing development of graphene oxide (GO)-based membrane technology for BL applications at Georgia Tech [2]. The fundamental methods for the synthesis and modification of these GO membranes will be illustrated, and the resulting separation properties such as permeation flux, lignin rejection, non-lignin organics rejection, inorganics rejections, and total solids rejection from BL feeds will be discussed in detail. The tuning of GO membranes to achieve nanofiltration (NF) and reverse osmosis (RO) performance will be demonstrated. Furthermore, the scale-up of GO membranes and their excellent performance in longer-term evaluation with realistic BL feeds (up to 3 gal/min) will be discussed. Finally, we will discuss the feasibility of a process that combines GO-based NF and RO membrane stages for concentration of BL as well as production of process-quality makeup water.

Background References:

[1] N. Kevlich, M. L. Shofner, S. Nair, "Membranes for Kraft Black Liquor Concentration and Chemical Recovery: Current Progress, Challenges, and Opportunities", *Separation Science & Technology*, 52 (6), p. 1070-1094 (2017).

[2] F. Rashidi, N. Kevlich, S. Sinquefield, M. Shofner, S. Nair, "Graphene Oxide Membranes in Extreme Operating Environments: Concentration of Kraft Black Liquor by Lignin Retention", *ACS Sustainable Chemistry & Engineering, 5,* p. 1002-1009 (2017).

Optimization Study of Microalgae Dewatering by utilizing Bio-based Alkali

Sheriff Olalekan Ajala[1*], Matthew L. Alexander[2]

[1]Sustainable Energy System Engineering, Frank H. Dotterweich College of Engineering, Texas A&M University-Kingsville

[2]Chemical and Natural Gas Engineering, Frank H. Dotterweich College of Engineering, Texas A&M University-Kingsville

*sheriffajalao@gmail.com

Abstract

In the present study, the flocculating viability of bio-based alkali derived from cocoa pod husk ash was investigated by using the D-optimal design of the response surface methodology (RSM). The interactive effects of biomass concentration and flocculant dose on flocculation efficiency and concentration factors were examined and the process was optimized using RSM. Under the optimum conditions of biomass concentration 0.54 g dry weight/L and flocculant dose 0.50 (% v/v), the maximum flocculation efficiency (92 %) and concentration factor (34) of *Dunaliella salina* were obtained. Meanwhile, the flocculation efficiency (95 %) and concentration factor (7) of *Nannochloropsis oculata* were obtained under the optimum condition of 0.83 g dry weight/L and 0.5 (% v/v) for biomass concentration and flocculant dose respectively. The result of elemental analysis of the bio-based alkali by ICP-MS showed that the concentration of K^+ (51489 ppm) is highest, followed by Ca^{2+} (1450 ppm) and Na^+ (210 ppm). This undoubtedly shows the alkaline nature of the obtained bio-based alkali from cocoa pod husk ash that was employed as flocculant for harvesting microalgae. This study confirmed that produced cocoa pod derived alkali can be used to effectively and efficiently harvest microalgae from dilute microalgae culture medium.

Introduction

The recognition of microalgae as an attractive biomass feedstock for biofuels and production of some industrial chemicals [1] is owing to many advantages of microalgae biomass over the first- and second-generation biofuel feedstocks, such as corns and soybeans. The economic feasibility of microalgae biomass as feedstock is limited to value added products such as nutritional supplements, due to high energy intensity associated with the biomass separation [2], [3]. Thus, the selection of a suitable, efficient, low energy requirement and low-cost dewatering process is imperative for industrial scale of low value products like biofuel. Flocculation followed by gravity settling of microalgae flocs among other microalgae dewatering techniques such as centrifugation, membrane filtration, is considered most efficient, economical and technological viable dewatering process [2], [4], [5]. The modeling and optimization study of chemical and biochemical processes for improved efficiency is considered as an important stage in developmental process [6]. Akış and Özçimen [7] and Abdul et al. [8] employed response surface methodology (RSM) for modeling and optimizing microalgae flocculation using alkalis and seed powder of clearing nut, *Strychnos potatorum* respectively.

The disposal of several tons of agricultural wastes such as cocoa pod husks has been a challenge among local farmers. Therefore, the conversion of such wastes into useful products like bio-based alkali derived from ash residue of such solid waste has been suggested.

In this study, the feasibility of bio-based alkali (BBA) derived from cocoa pod husk to induce flocculation of *Dunaliella salina* and *Nannochloropsis oculata* was investigated. Also, the flocculation process variables, biomass concentration and flocculant dose, were optimized by using RSM. Biomass flocculation efficiency and concentration factor were considered as response variables.

Experimental

Materials. The waste cocoa pod husks used for this work were obtained from in Nigeria. Thermolyne benchtop muffle furnace was used for calcinating the cocoa pod husk. The microalgae species (*Nannochloropsis oculata* and *Dunaliella salina*) used were bought from UTEX Culture Collection of Algae at The University of Texas at Austin. All laboratory reagents utilized are of analytical grade purchased from Fisher Scientific International, Inc. and VWR International Inc.

Microalgae cultivation. The *N. oculata*, and *D. salina* were cultured for 21 days in 1.5-liter photobioreactors, The Artificial Seawater and 2X Erdschreiber's growth media were respectively used. The culture was air sparged at 0.02 vvm with 2 % CO_2 sparging rate, and light was supplied by using daylight florescent tubes at 3500 lux with 12:12 h light-dark cycle. The flocculation experiment was carried out after 21 days of culturing with microalgae biomass concentration of 816 and 825 mg/L for *N. oculata*, and *D. salina*, respectively.

Preparation of BBA and characterization. The cocoa pod husks were washed several times using deionized water. The cocoa pod husks were cut into smaller pieces and sun dried to constant weight. Thereafter, the dried cocoa pod husk is calcinated in a muffle furnace at 540^0C for 5 hours.

The ash obtained was maximally extracted (at three different concentrations) for 48 hrs. using distilled water. Thereafter, the mixtures were filtered with Whatman filter papers to remove the remaining solids. The pH of the filtrates was measured using an Orion Star A211 pH meter, and the filtrate with the highest pH was chosen for the flocculation experiment. Also, the filtrate and the cocoa pod husk ash were characterized by inductively coupled plasma mass spectrometry (ICP-MS) and scanning electron microscope-energy-dispersive X-ray spectroscopy (SEM-EDS), respectively.

Experimental design for flocculation and RSM modeling. A three-level-two factor D-optimal design was employed in modeling and optimizing the flocculation of microalgae. About 16 experimental runs were generated for each of the two species of microalgae. The independent variables considered include biomass concentration (X_1) and flocculant dose (X_2), while the flocculation efficiency (Y_1) and concentration factor (Y_2) were the dependent variables. Table 1 illustrates the coded level of process parameters studied in this work. In order to relate the response variables to the two independent variables employed, multiple regressions were used to fit the coefficient of the second order mathematical model of the responses. The goodness of fit of the model was evaluated by using a test of significance and an analysis of variance (ANOVA). Equation 1 presents the general form of the fitted second order mathematical response of the model.

$$Y = \alpha_0 + \sum_{i=1}^{n} \alpha_i X_i + \sum_{i=1}^{n} \alpha_{ii} X_i^2 + \sum_{i<j}^{n} \alpha_{ij} X_i X_j + \varepsilon \quad (1)$$

Y is the predicted response (flocculation efficiency or concentration factor), α_0 is the intercept, α_i, α_{ii}, $\propto_{ij}$ (i = 1, 2, ..., n) are the linear, second order and interaction coefficients, respectively. ε is the error term of the model.

Flocculation Experiment. The effects of flocculating parameters such as biomass concentration and BBA dosage were investigated by analyzing the biomass recovery and concentration factor. About 20 ml of microalgae suspension was placed in a 50 ml beaker, and the prepared BBA was added, depending on required dosage stated in the experimental design, while the mixture was stirred with a magnetic stirrer at 80 rpm for 2 minutes, then followed by gentle mixing at 10 rpm for 20 minutes. Thereafter, the mixture was poured into a graduated cylinder. The suspension was allowed to settle for 30 minutes and a sample was taken from the middle of the clarified zone. The optical density of the samples was measured with a Unico S1200 Visible spectrophotometer. Also, the initial height of the microalgae suspension and the settled floc height were measured using a meter rule. These procedures were repeated in duplicate for *N. oculata* and *D. salina*. Thus, the

Table 1
The D-optimal design ranges and level of independent parameters studied with (a) *D. salina* and (b) *N. oculata*.

(a)		Coded factor level		
Factor	unit	-1	0	+1
Biomass Concentration, X_1	g/L	0.494	0.916	1.767
Flocculant dose, X_2	% v/v	0.500	1.000	2.000
(b)		**Coded factor level**		
Factor	unit	-1	0	1
Biomass Concentration, X_1	g/L	0.210	0.420	1.080
Flocculant dose, X_2	% v/v	0.500	1.000	2.000

biomass recovery and concentration factor were calculated as follows;

$$\text{Biomass recovery}, \eta = \frac{OD_{650i} - OD_{650f}}{OD_{650i}} \quad (2)$$

$$\text{Concentration factor} = \frac{H_0}{H_f} \quad (3)$$

The variable $OD_{6\ 5\ 0}$$_{i}$= optical density of microalgae suspension was measured at 650 nm before the addition of BBA,$OD_{6\ 5\ 0}$$_{f}$= optical density was measured at 650 nm, after the addition of BBA and settling. The H_0 and H_f are the initial microalgae suspension and the settled floc heights respectively.

Results and Discussion

Tables 2 and 3 show the elemental composition of ash sample and filtrate with potassium having the highest concentration of 51,489 ppm, which corresponds to the highest peak in the chromatogram in Fig. 1a. The SEM image of cocoa ash is shown in Fig. 1b.

The RSM optimization results indicated biomass concentration of 0.54 g/L and 0.50 % v/v flocculant (BBA) dose for the *D. salina* and 0.50 g/L biomass concentration, 0.50 % v/v flocculant (BBA) dose for *N. oculata*. The RSM values predicted at the optimum conditions were flocculation efficiencies of 92 and 95 % and the concentration factors of 34 and 7 for *D. salina* and *N. oculata*, respectively. The optimum predicted response variables by RSM was validated experimentally using two independent replicates. The experimental values obtained at optimum conditions were flocculation efficiency (%) and concentration factor of 91.37±1.27 and 32.48±0.59, respectively for *D. salina*. The *N. oculata* flocculation efficiency (%) and concentration factor were 94.61±0.58 and 8.55±0.67, respectively at the optimum conditions. The results obtained confirmed the efficacy of the produced bio-based alkali from waste cocoa pod husk, which can serve as flocculant to induce microalgae flocculation.

Table 2
Elemental composition and concentration of bio-based alkali, BBA and ash.

Element	Weight percent†	Concentration (ppm)‡
Carbon, C	11.56±1.36	NA
Oxygen, O	43.72±2.33	NA
Sodium, Na	ND	310.19
Magnesium, Mg	2.28±0.28	2.69
Aluminum, Al	ND	ND
Silicon, Si	0.34±0.04	NA
Phosphorus, P	1.29±0.04	NA
Sulfur, S	1.09±0.15	NA
Chlorine, Cl	.20±0.01	NA
Potassium, K	37.24±4.72	51489.27
Calcium, Ca	2.31±0.54	1450.10
Manganese, Mn	ND	11.71
Iron, Fe	ND	5.95

†- EDS values, ‡- ICP-MS values, NA-not applicable, ND-not detected

Table 3
Physiochemical properties of BBA.

Property	Values
pH[a]	10.55
pH[b]	11.53
pH[c]	12.37
Concentration (M)	0.347
Color	colorless
Conductivity (mS)	19.99
Temperature (°C)	20°C
TDS[d]	1.28×10^4

pH[a,b,c]- pH of samples concentration of 0.075, 0.100, 0.125g/mL, TDS[d] - total dissolved solids

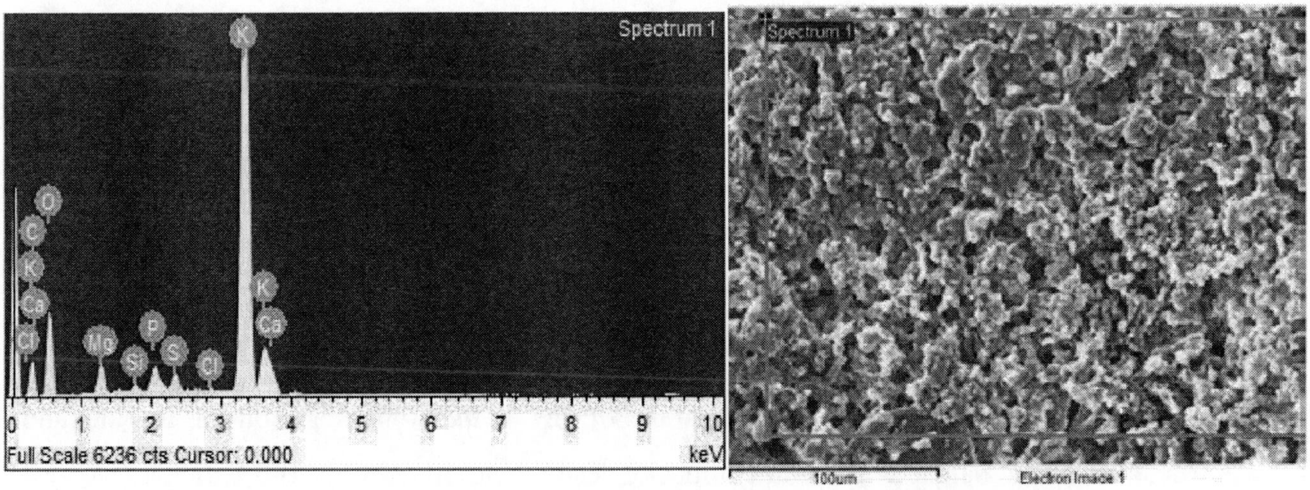

Fig. 2. Elemental spectra by SEM-EDS (a) SEM image (b) of the cocoa pod ash.

Conclusion

This present study confirmed the use of BBA that can possibly supplant commercial alkali in alkaline flocculation process of microalgae. This can possibly lower the microalgae harvesting cost. The application of BBA to induce microalgae flocculation resulted in more than 90% flocculation efficiency at the optimum conditions. The prospect of BBA from cocoa pod husk to induce flocculation in microalgae culture is credited to its high potassium content.

References

[1] D. Vandamme, S. Eyley, G. Van den Mooter, K. Muylaert, and W. Thielemans, "Highly charged cellulose-based nanocrystals as flocculants for harvesting Chlorella vulgaris," *Bioresour. Technol.*, vol. 194, pp. 270–275, 2015.

[2] E. S. Salama *et al.*, "Harvesting of freshwater microalgae Scenedesmus obliquus and Chlorella vulgaris using acid mine drainage as a cost effective flocculant for biofuel production," *Energy Convers. Manag.*, vol. 121, pp. 105–112, 2016.

[3] M. Bayat, R. Khalilzadeh, and H. Kouchakzadeh, "Efficient harvesting of marine Chlorella vulgaris microalgae utilizing cationic starch nanoparticles by response surface methodology," *Bioresour. Technol.*, vol. 243, pp. 583–588, 2017.

[4] T. Chatsungnoen and Y. Chisti, "Harvesting microalgae by flocculation-sedimentation," *Algal Res.*, vol. 13, pp. 271–283, 2016.

[5] S. Lama, K. Muylaert, T. B. Karki, I. Foubert, R. K. Henderson, and D. Vandamme, "Flocculation properties of several microalgae and a cyanobacterium species during ferric chloride, chitosan and alkaline flocculation," *Bioresour. Technol.*, vol. 220, pp. 464–470, 2016.

[6] D. Bas, "Modeling and optimization II : Comparison of estimation capabilities of response surface methodology with artificial neural networks in a biochemical reaction," vol. 78, pp. 846–854, 2007.

[7] S. Akış and D. Özçimen, "Optimization of pH induced flocculation of marine and freshwater microalgae via central composite design," no. March, pp. 1–6, 2019.

[8] S. Abdul, S. Duraiarasan, and A. S. S. Shellomith, "Statistical optimization of harvesting Chlorella vulgaris using a novel," *Biotechnol. Reports*, vol. 7, pp. 150–156, 2015.

581377 Renewable Deep Eutectic Solvents (DES): Making Possible the Biochemical Conversion of Lignin

Colombia has a vast plantation of sugar cane and one of the highest sugar productivities of the world (Ton per hectare) what causes an annual production of more than 6.5 million tons of bagasse on the region. Sugar cane bagasse is approximately composed of 45,3 % cellulose, 27,5% hemicellulose and 23,6% lignin. However, lignin solubilization and extraction has been recognized by the scientific community as the main limitation for its efficient conversion. Existing methods cannot recover a large quantity of high-purity lignin without alterating its chemical structure.

This study reports a new method to extract and dissolve lignin from sugar cane bagasse produced in Valle del Cauca-Colombia region using a set of in-house synthesized Deep Eutectic Solvents (DES). The DES is a mixture of two or more chemicals acting as either hydrogen-bond donors (HBD) or hydrogen-bond acceptors (HBA), and it has been recently reported as a potential solvent for lignin extraction. We design the DESs to use only renewable materials that can be obtain from biomass. The resulting DES lignin has unique structural features with high potential towards a biocatalytic conversion approach. The efficient extraction of lignin and the increase of its reactivity towards biological conversion is due to the exposition of the new active sites for conversion, as lignin conformational changes from a rigid structure to a relaxed conformation during solubilization.

This study presents an opportunity to find a set of suitable conditions to facilitate extraction and solubilization of high purity lignin from sugar cane bagasse with high purity and minimal chemical modifications, which will help establish breakthrough pathways in lignin Biochemical valorization research.

581380 Purification of Lignocellulosic Hydrolyzate using Flocculant enhanced Depth Filters

For lignocellulosic biorefineries to become economically viable, efficient separation processes must be developed. The removal of lignin-based contaminants prior to fermentation is a major bottleneck for hydrolysate valorization. Fibrous depth filtration is commonly used to remove particles from suspension. Traditional depth filters are ineffective at removing colloidal impurities from hydrolysate streams. To enhance removal efficiency, flocculating polymers are added upstream to aggregate oppositely charged particles. In this experiment, these two operations are integrated. Cationic flocculant poly (diallyldimethylammonium chloride) (PDADMAC) was adsorbed onto the surface of cellulose pulp fibers. The effects of polymer treatment dosage and ratio the of cellulose to PDADMAC were examined. Dynamic light scattering shows the effect on particle size. Zeta potential measurements yield the stability of the suspension. UV/VIS quantified the removal of lignin from the suspension onto the fibers:flocculant complex. The results showed that PDADMAC adsorbed pulp fibers can effectively remove contaminants from aqueous suspensions.

558654 Subcritical CO_2-Assisted Autohydrolysis for the Co-Production of Oligosaccharides and Fermentable Sugar from Corn Straw

Subcritical CO_2-assisted autohydrolysis pretreatment for oligosaccharides and fermentable sugars production from corn straw was investigated. The result showed that the prehydrolysate was rich in xylose, and about 90.22% were existed in the form of xylo-oligosaccharides (XOS), as well as the functional XOS (DP<5) took up about 40%. Few inhibitors were formed during the pretreatment and almost no effect on subsequent fermentation. Additionally, the highest glucose yield was 90.59% (32.84g/100g corn straws) and 89.44% (21.17g/100g corn straws) of lignin was recovered in the enzymatic hydrolysis residue, it can be concluded that the proposed pretreatment method was an effectively way to improve corn straw enzymatic hydrolysis with few lignin degradation. Furthermore, the characterization data were also obtained through the analysis of pretreated solids at 170°C, 5Mpa CO_2 and 40min residence time using HSQC, XRD, FTIR, SEM and TG analyses.

564784 Cell Wall Disruption in Low Temperature By KOH/Urea and Potential Application of Its Liquid Waste in Crop Production

Cell wall disruption in low temperature by KOH/urea and potential application of its liquid waste in crop production

Xinshu Zhuang, Zahoor, Wen Wang, Xuesong Tan, Qiang Yu, Zhenhong Yuan

Guangzhou Institute of Energy Conversion, Chinese Academy of Sciences; CAS Key Laboratory of Renewable Energy; Guangdong Provincial Key Laboratory of New and Renewable Energy Research and Development, Guangzhou 510640, P. R. China

Abstract

Pretreatment of corn straw by KOH/Urea (KU) aqueous solution at low temperature was investigated. To study the cell wall distraction during this low temperature process, and its impacts on enzymatic hydrolysis, cellulose crystallinity, and chemical properties were investigated by High performance liquid chromatography, Scanning electron microscopy, Fourier transform infrared spectra, and X-ray diffraction. The results showed that the degradation of plant cell wall at low temperature was attributed to distraction of the hydrogen bonds in cellulose and solubilization of hemicellulose and lignin. Meanwhile, the pretreatment approach resulted in increasing cellulose recovery up to 29%, approximately 33% of lignin and 15% of hemicellulose removal, respectively. Under KU pretreatments, biomass samples exhibited much enhanced hexoses yields from enzymatic hydrolysis, a higher saccharification (80%) of cellulose in pretreated materials was achieved with domestic cellulase loading of 20 FPU/g substrates within 2h at 30 ^{0}C. Further co-supplied with 1% Tween-80 exhibited an almost complete enzymatic saccharification of the residues. It is interesting to note that the pretreatment liquid waste (PLW) were applied as phytogenic molecular activator, leading to increase dry biomass yields by 3¨C24 times compare to control in different crops. Hence, this study has not only demonstrated a new pretreatment technology for a complete saccharification, but it has also produce phytogenic molecular activator for the crop growth through PLW without any negative impact on environment, which laying the foundation for the cellulosic ethanol bio-refinery.

Keywords: Pretreatment, Lignocellulose, KOH/Urea, Cell wall disruption, liquid waste, phytogenic molecular activator, crop production.

This work was financially supported by the National Key Research and Development Program of China (2018YFB1501402), the National Natural Science Foundation of China (51876206,51606203,51476179).

564908 Lignocellulosic Biomass Derived N and S Doped Porous Carbon for Efficient Oxygen Reduction Reaction

The oxygen reduction reaction (ORR) represents a key cathode reaction in fuel cells and metal–air batteries but high cost due to noble metal as catalyst limits its large-scale implementation. Lignocellulosic biomass is potential feedstock in carbonization for sustainable, active, inexpensive, and stable electrocatalyst preparation to replace Pt-based materials for the oxygen reduction reaction. Heteroatom doping (N, B, S, et al) can enhance carbon materials' electrocatalyst performance by improving the wettability and conductivity. Lignosulfonate (LS) is a cheap and common biomass and contain S and inspired by this, we dope N into LS to obtain N and S co-doped porous carbon by ammoxidation and carbonization and used as electrocatalyst in ORR approaching the performance of 20 wt% Pt/C. Furthermore, we are conducting works on modification of technical lignins followed by heteroatom doping and carbonization and nonprecious metals loading for ORR and more electro-chemistry for example Hydrogen and Oxygen Evolution Reactions (HER and OER).

573441 Oxidation and Depolymerization of Alpha,Gamma-Diol-Protected Lignin into Aromatic Monomers

Lignin oxidation offers a potential pathway to sustainable oxygenated aromatic molecules. However, current methods that use real lignin tend to have low selectivity and a yield that is limited by lignin degradation during its extraction. Here, we show that a,γ-diol protected lignin, which can be isolated from birch at near quantitative yields using propionaldehyde under acidic conditions, can be upgraded at high selectivities (>90%) to just two products. We developed stoichiometric and catalytic oxidation methods using 2,3-dichloro-5,6-dicyano-1,4-benzoquinone (DDQ) as oxidant/catalyst to selectively deprotect the acetal and oxidize the a-OH into a ketone. The oxidized lignin was then depolymerized using a formic acid/sodium formate system to produce aromatic monomers with a 38 mol% (in the case of stoichiometric oxidation) and 32 mol% (in the case of catalytic oxidation) yield (based on the extracted lignin). The monomer products contained 90-93% of the syringyl or guaiacyl propane diones and 7-10% of phenyl acids and aldehydes. These unprecedented selectivities are attributed to the preservation of the lignin structure by the acetal. Taking the high lignin isolation yield into account, this method also produced higher overall aromatic monomer yields (36 mol% for stoichiometric oxidation and 31 mol% for catalytic oxidation based on the original Klason lignin) compared to other lignin oxidation and depolymerization strategies.

567563 Hybrid Dilute Sulfuric Acid and Aqueous Ammonia Pretreatment for Improving Butanol Production from Corn Stover with Reduced Wastewater Generation

An efficient hybrid pretreatment method was developed for butanol production from corn stover using dilute sulfuric acid (DA) and aqueous ammonia (AA). With the optimized AA concentration, treatment temperature and time of 10% AA, 80 degrees C and 24 h, the hybrid pretreatment could effectively dissolve hemicellulose and lignin with solid recovery rate of 37.45% and lignin reduction rate of 86.77% compared to those of 57.75% and 45.84% from single DA pretreatment. By washing 1 time after each step treatment, sugar yield and butanol production were increased to 401.76 mg/g-CS and 10.89 g/L from 346.04 mg/g-CS and 9.33 g/L obtained without washing. Compared with conventional single DA and AA pretreatment methods, wastewater generation was reduced to 0.83 L/g-butanol from 2.11 and 3.46 L/g-butanol, indicating this hybrid pretreatment could be an effective approach for improving butanol production from lignocellulosic feedstocks.

568657 Kinetic Study of Dissolution of Biochar in Tetralin

The need for renewable energy across the world has recently increased due to the effects of global warming. One of the primary alternatives that researchers are investigating, in order to replace or at least complement fossil fuels and mitigate the effects of global warming, is the production of biofuels from biomass. Biomass consists of 30-35 wt% cellulose, 15-35 wt% hemicellulose, and 20-35 wt% lignin. In our research, biochar was first produced from cellulose and lignin by hydrothermal carbonization (HTC). In HTC, each biomass model compound (cellulose or lignin) was mixed with water and heated to 220C-370C in a closed batch reactor to form biochar, a carbon rich solid. In the second step, the dissolution of cellulose and lignin biochar in a hydrogen donor solvent (tetralin) was studied at 400°C (temperature at which tetralin breaks down to naphthalene and hydrogen) for 60 mins. This liquefaction process, analogous to the direct coal liquefaction, has the potential to produce value-added chemicals like aromatics. In order to understand the effect of temperature on product distribution, the dissolution experiments were stopped at various points (200°C, 250°C, 300°C, 350°C and 400°C) during the heating stage. For the isothermal stage at 400°C, the reaction was stopped after 15, 30, 45 and 60 mins to investigate the kinetics of the reaction. The results would provide a detailed look into the conversion of biochar and yield of the different products formed at various stages of the dissolution reaction, crucial for process design. The products formed during dissolution at various stages were analyzed using gas chromatography-mass spectroscopy and mainly consisted of pre-asphaltenes (tetrahydrofuran soluble), asphaltenes (toluene soluble) and oil (hexane soluble).

578280 Techno-Economic Analysis of Production of Mixed-Furans from Biomass Hydrolysate

Lignocellulosic biomass is a sustainable feedstock for production of furans – 5-hydroxymethy furfural (HMF) and furfural – versatile platform molecules that can be converted to gasoline additives and bio-based polymers. Furans can also be used to synthesize wide-array of industrially-important molecules like γ-valerolactone, 1,6-hexanediol, adipic acid, tetrahydrofuran, levulinic acid, etc. Noteworthy among these is furandicarboxylic acid – a precursor for polyethylene furanoate that can replace fossil-derived polyethylene terephthalate. We present a techno-economic analysis of a process for high-yield production of mixed-furans (5-hydroxymethyfurfural and furfural) using the recently developed Simultaneous-Isomerization-Reactive-Extraction and Back-Extraction (SIRE-BE) process followed by dehydration.[1]

In the SIRE-BE-Dehydration process, glucose and xylose are first isomerized to corresponding ketose forms (i.e. fructose and xylulose) that are selectively extracted into an immiscible organic phase. Next, the keto-sugars are back extracted from the organic phase into an acidic water medium. Keto-sugars undergoes dehydration reaction in presence of acetone solvent, which is uniquely suited for high-yield of furans by solvation. Finally, the formed furans and acetone are extracted into non-polar toluene and simultaneously separation of water is achieved. We simulated the steady state operation of a SIRE-BE-Dehydration process in a bio-refinery with feed rate of 1000 MT of mixed-sugars per day. Using discounted cash flow analysis, the Minimum Furan Selling Price (MFSP) was estimated to be $1.50 /kg – a promising price-point for industrially-important chemical products. Sensitivity analyses identified that the overall yield and feedstock sugar costs as key factors affecting the Minimum Furan Selling Price. With reasonable pathway to achieve a price which is competitive to be used for fuel application and polymer precursors will be discussed.

[1] Alipour, S.; Relue, P. A.; Viamajala, S.; Varanasi, S., High yield 5-(hydroxymethyl)furfural production from biomass sugars under facile reaction conditions: a hybrid enzyme- and chemo-catalytic technology. *Green Chem.* **2017,** *19* (7), 1782.

576332 Oxidative Degradation of Lignin to Value Added Products

Kraft lignin obtained from paper and biofuel productions is currently being used as energy, however it can be further converted to value added products thereby making sustainable bio-refinery. Oxidative degradation of lignin produces low molecular weight phenolic compounds which are high value added chemicals. However, the current challenge of this process is to selectively control and improve the yield of these phenolic compounds. In addition to phenolic compounds, dicarboxylic acids are also produced by fragmentation of aromatic ring of lignin. The goal of this research is to deepen the fundamental understanding of factors governing the oxidative degradation processes of kraft lignin and integrate it to current pulp and paper industry. This will be done by experimental study of different reaction conditions, oxidizing and stabilizing agents, lignin source and reactor configurations. The products were detected using high performance liquid chromatography using RI and UV detectors and verified with gas chromatography mass spectrometry. This approach is focused on analyzing factors improving the yield of low molecular weight phenolic compounds. Hydrogen peroxide is of the most widely used bleaching agent in the pulp and paper industry. Hence, the use of hydrogen peroxide as an oxidizing agent improves the economic viability of integrating the process to pulp and paper industry. The study conducted showed that hydrogen peroxide treatment of kraft lignin can be carried out at a comparatively low severity of reaction conditions compared to oxygen. The preliminary results indicate that experiments carried out under certain conditions generate low molecular weight phenolic compounds in yields of 1-2 % and carboxylic acid fragments in yields of 15-20%. These compounds are essentially phenolic aromatic aldehydes such as vanillin, syringaldehyde, vanillic acid and carboxylic acids such as oxalic and formic acids. These preliminary results indicate that there is a scope of improving low molecular weight phenolic aldehydes and carboxylic acids yield. Also, the techno-economic feasibility of the process makes it suitable to further deepen the fundamental understanding of factors governing the peroxide degradation processes. Lignin model compound study will be performed under similar conditions to further develop the reaction scheme. This study will also include addition of aromatic ring preserving agents such as a-tocopherol. Finally, the method to selectively generate low molecular weight phenolic compounds from kraft lignin will be proposed. The resultant process then will be integrated to pulp and paper industry which will help in diversifying the products portfolio for a sustainable bio-refinery.

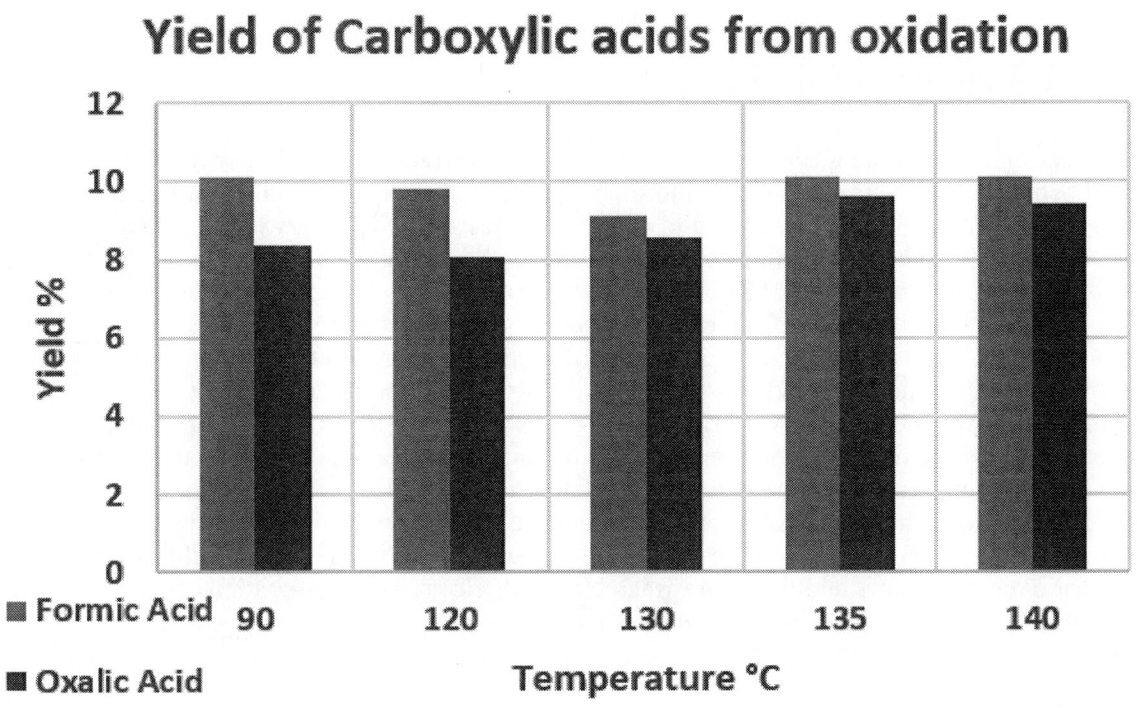

Figure 1. Yield of carboxylic acids from oxidation of lignin

575452 Quantitation, Depolymerization, and Demethylation of Lignin in Inorganic Ionic Liquid

It was found that lignin could be quantitatively isolated as a solid fraction from lignocellulose when cellulose and hemicellulose were hydrolyzed into monosugars in an inorganic ionic liquid (acidic lithium bromide trihydrate (ALBTH) system) under mild conditions. Based on the observations, a facile and fast method was developed for the quantitation of lignin in biomass, which gave comparable results with the existing Klason (sulfuric acid) method. The characterization of the isolated lignin indicated that the isolated ALBTH lignin had high purity, low molecular weight, and non-condensed moieties (i.e., Hibbert's ketones and benzodioxanes), which had not been identified and detected in other isolated lignins before. Study using lignin model compounds revealed that the lignin depolymerization was catalyzed by LiBr and acid synergistically via α-benzyl carbocation intermediate. The cleavage of the β–*O*–4-aryl ether bond led to the Hibbert's ketone, while the ring formation between the α-benzyl carbocation and the methoxyl group on the adjacent benzene ring resulted in the benzodioxane structures. Condensation of lignin model compounds occurred when treated in ALBTH system, but the condensation of the real lignin in biomass was very limited in the ALBTH, because the limited and blocked mobility and accessibility of both the a-benzyl carbocation intermediates and the lignin fragments with electron-rich aromatic sites in solid state prevented them from condensation. Further investigation indicated that lignin could be selectively demethylated in the ALBTH system.

573878 Oxidation of Lignin-Rich Residue from Deacetylation, Mechanical Refining, and Enzymatic Hydrolysis of Lignocellulose

Production of aromatic aldehydes, such as vanillin, from lignin, has been an intriguing process for decades, but widespread implementation has been inhibited by the high hydroxide:lignin ratios required for significant aldehyde yields, as well as the widely different chemical structure of industrial lignins. In this work, we explore oxidation of the lignin-rich residue isolated from corn stover after deacetylation, mechanical refining, and enzymatic hydrolysis (DMR lignin), including a survey of reaction conditions and detailed product characterization that accounts for greater than 60% of substrate carbon as quantified products (aromatic monomers, aliphatic acids, and carbonate) and the remainder as an oligomeric oil. We also discuss strategies for maximizing selectivity to desired products and separating them from the alkaline solution.

573265 Quantum Mechanical Studies of the Depolymerization of Lignin By the $ZnCl_2$:Ethylene Glycol Deep Eutectic Solvent

Lignin makes between 15-35 % of the worlds dry biomass. It consists of phenolics connected to each other through various linkages. In order to convert lignin into value added chemicals, the phenolic-phenolic linkages must be broken, a process known as depolymerization. Several methods, such as pyrolysis, catalytic reductive or oxidative, and the use of ionic liquids (IL), have been reasonably effective, but are fraught with undesired toxic reagents, high costs, or modest yields. Compared to ILs, Deep Eutectic Solvents (DES) are lower in cost, easily prepared, and efficiently recycled. DESs have found utility in several areas, but in the realm of lignin chemistry, have been shown to solvate lignin. In addition, several metal chlorides (MCl), such as $ZnCl_2$, and $AlCl_3$, have been shown to aid in the depolymerization of lignin. In the combination of a metal chloride (MCl), a DES can be formed which both solvates lignin, and also aids in its depolymerization. Indeed, one such DES consisting of ethylene glycol, and $ZnCl_2$ in a 4 to 1 molar ratio has been developed and shown to solvate lignin in concentrations of at least 10 Mol %. One area, which still needs to be understood, is the mechanism by which $ZnCl_2$ aids depolymerization. To that end, Quantum Mechanics calculations were carried out on a series of tri-lignol + $ZnCl_2$ Potential Energy Surfaces (PES). Tri-lignols are amenable to high level Quantum Mechanics calculations, yet are complex enough to shed insight into the depolymerization process. Our results show that there are several places on lignin which $ZnCl_2$ can bind to. Of these, the strongest attractions occur between zinc and aromatic carbons, as well as to the oxygen atom of a ß-O4 linkage. Our results add to the knowledge of catalysis in lignin chemistry, and also serve as a foundation for the development of future lignin fractionation strategies.

570147 Non-Catalytic Oxidative Depolymerization of Lignin with Molecular Oxygen to Produce Valuable Phenolic Monomers

Lignin as the only component of lignocellulosic biomass containing aromatics has gained increasing attention over the past decades. Lignin valorization could improve the efficiency and profitability of lignocellulosic biorefineries significantly. Recently, oxidative depolymerization of lignin into aromatic compounds have received considerable attention. Unlike hydrogenolysis, oxidative conversion does not require expensive hydrogen and can produce valuable phenolic monomers like vanillin and syringaldehyde. Our hypothesis was that at sufficiently high temperatures, lignin decomposition generates radicals that can activate molecular oxygen without any catalyst. Moreover, activated molecular oxygen can break C-O and C-C bonds, enhancing lignin decomposition and generating additional radicals. In order to test our hypothesis, technical and native lignin sources were depolymerized in presence of oxygen in various polar aprotic and polar protic solvents including acetonitrile, tetrahydrufuran, ethanol and a nonflammable solvent with high oxygen solubility. The reaction products were analyzed through GC/MS, GPC and 2D HSQC NMR techniques. Gas chromatogram of silylated bio-oil showed production of vanillin, syringaldehyde, acetosyringone, syringic acid and vanillic acid as the major monophenolic compounds. GPC results of the obtained bio-oil under oxygen atmosphere also showed lower molecular weight of the bio-oil compared to the obtained bio-oil under inert atmosphere which proves the need for oxygen to depolymerize lignin.

567010 Lignin Bioconversion: Valorization from Heterogeneous Stream to High-Value Target Compound

The heterogeneity of lignin is one of the major challenges for the efficient utilization and commercial valorization of lignocellulosic biomass. Without fully utilization, many lignin streams were burned or even discharged directly, which not only causes waste of resource, but also causes many serious environment problems. The produced aromatic compounds were often heterogeneous streams, for the intrinsic heterogeneity of lignin, which in turn present significant technical challenges when attempting to produce lignin-derived chemicals where purity is often a concern. To overcome this challenge, a microbial bioconversion system will be established by fully using the capacity of the microbe in aromatic compounds catabolism in a funneling-like approach. With the tools of synthetic biology and metabolic engineering, a lignin bioconversion system to produce target compound was synthesized in the bacteria by integrating the lignin depolymerization module and target compound biosynthesis module. A laccase and versatile peroxidase co-secretive expression system was introduced to the bacteria for lignin depolymerization to release aromatic monomers. Chemical structure characterization and bioinformatic analysis were carried out to understand the mechanism of lignin degradation in the bacteria, and further guide the biodesign and optimization of product biosynthesis module. Multiple strategies were applied to optimize the target compound biosynthesis pathway, including replacing with efficient enzyme, promoter or RBS to relieve the bottleneck step, and knocking out the critical genes to block the competitive inhibition pathways. Overall, the study will enable an efficient lignin conversion to target compound with a "one step" microbial fermentation. The outcomes of this study will provide valuable fundamental information for lignin bioconversion to produce high-value aromatic compounds, as well as for the other chemicals production, and thus will promote the developing of industrial valorization of lignocellulosic biomass.

577604 Analysis of Pyrolysis Oil Model Compounds Using *in-Situ* raman

In recent decades, the world has seen an exponential increase in the demand for fossil-based energy sources. This increase in the demand is leading to an exhaustion of the petroleum reserves and increase in the environmental pollution. Therefore, attention is being given to alternative, carbon-neutral renewable energy sources like biofuels. The carbon-neutrality and lower production costs make the production of biofuel from lignocellulosic biomass an alluring alternative to conventional fossil fuels. Biofuels are generally produced from biomass via pyrolysis which yields pyrolysis oil that is high in oxygenated compound. The pyrolysis oil is then upgraded to biofuel through hydrodeoxygenation (HDO). Analysis of the HDO reactions for conversion and kinetic parameters often requires gas chromatography (GC) with time-consuming material preparation steps prior to the GC analysis. An alternate, fast analytical method would be beneficial to both researchers and producers alike.

Therefore, in this research, we propose the use of *in-situ* Raman Spectroscopy which can yield us time-resolved information about molecular structure, intermediates, mechanisms, and rate of reaction. The HDO reaction was simulated by analyzing multiple samples of varying concentrations of the reactants, and their corresponding HDO products. The obtained Raman spectra was then deconvoluted using Gaussian Deconvolution algorithm, in order to identify, and differentiate between different peaks. The height of their respective characteristic stretching modes were then measured to determine the conversion of the model compound. The carbonyl ($\nu_{C=O}$) stretching mode in hydroxyacetone, for instance, was found at 1,721 cm^{-1}, (ν_{COCOC}) sym. stretching mode of the acetal group found in glycolaldehyde diethyl acetal was observed at 1,140 cm^{-1} and the phenol (ν_{C-O}) stretching mode of 4-n-propylphenol was obtained at 1,204 cm^{-1}, respectively. A calibration curve was then obtained from the corresponding peak height versus concentration. Insofar, *in-situ* Raman could be used to measure the conversion of the hydrodeoxygenation of pyrolysis oil compounds.

AUTHOR INDEX

AUTHOR INDEX

AUTHOR INDEX

AIChE
120 Wall Street, FL 23
New York, NY 10005-4020

ISBN 978-1-7138-0539-7

9 781713 805397

CURRAN ASSOCIATES INC.
proceedings.com